SEMILINEAR ELLIPTIC EQUATIONS FOR BEGINNERS

Semilinear Elliptic Equations for Beginners

Editor

Qing Jun Hou

scitus
academics

Semilinear Elliptic Equations for Beginners
Edited by **Qing Jun Hou**

ISBN: 978-1-68117-569-0
Library of Congress Control Number: 2016943001

SCITUS Academics LLC,
www.scitusacademics.com
Box No. 4766, 616 Corporate Way,
Suite 2, Valley Cottage,
NY 10989

Notice

Printed in United States of America on Acid Free Paper

PREFACE

Elliptic equation is a class of partial differential equations describing phenomena that do not change from moment to moment, as when a flow of heat or fluid takes place within a medium with no accumulations. The Laplace equation, $u_{xx} + u_{yy} = 0$, is the simplest such equation describing this condition in two dimensions. In addition to satisfying a differential equation within the region, the elliptic equation is also determined by its values (boundary values) along the boundary of the region, which represent the effect from outside the region. Semilinear elliptic equations play an important role in many areas of mathematics and its applications to other sciences. Semilinear elliptic equations are of fundamental importance for the study of geometry, physics, mechanics, engineering and life sciences. The variational approach to these equations has experienced spectacular success in recent years, reaching a high level of complexity and refinement, with a multitude of applications.

Semilinear Elliptic Equations for Beginners is a comprehensive guide to variational methods and their applications to semilinear elliptic problems. This book deals with nonlinear boundary value problems for semilinear elliptic equations on unbounded domains. This book will be of valuable for professors, practitioners, and researchers in mathematics and mathematical physics.

TABLE OF CONTENTS

1

MULTIPLE SOLUTIONS FOR A CLASS OF SEMILINEAR ELLIPTIC EQUATIONS WITH NONLINEAR BOUNDARY CONDITIONS

Ziyan Yao

Department of Mathematics, University of Shanghai for Science and Technology, Shanghai, China

ABSTRACT

In this paper, using Local Linking Theorem, we obtain the existence of multiple solutions for a class of semilinear elliptic equations with nonlinear boundary conditions, in which the nonlinearites are compared with higher Neumann eigenvalue and the first Steklov eigenvalue.

KEYWORDS

Multiple Solutions; Nonlinear Boundary Conditions; Local Linking Theorem

1. INTRODUCTION

In this paper, we investigate the multiple solutions for semilinear elliptic equation with nonlinear boundary conditions

$$\begin{cases} -\Delta u + c(x)u = f(x,u), x \in \Omega \\ \dfrac{\partial u}{\partial v} = g(x,u), \qquad x \in \partial\Omega \end{cases} \tag{1.1}$$

where $\Omega \subset \mathbb{R}^N (N \geq 3)$ is bounded domain with smooth boundary $\partial\Omega$ and $\frac{\partial}{\partial v} = v \cdot \nabla$ is the outward normal derivative on $\partial\Omega$, and the function $c(x)$ satisfies C) $c(x) \in L^\infty(\Omega)$, and $c(x) \geq 0$ a.e. on Ω with $\int_\Omega c(x)dx > 0.$

Problems of the above type have been discussed extensively. In 1902, Steklov (see [1]) studied the eigenvalue problem

$$\begin{cases} -\Delta u = 0, x \in \Omega \\ \dfrac{\partial u}{\partial v} = \lambda u, x \in \partial\Omega \end{cases} \tag{1.2}$$

Auchmuty (see [2]) considered the eigenvalue problem

$$\begin{cases} -\Delta u(x) + c(x)u = 0, x \in \Omega \\ \dfrac{\partial u}{\partial v} = \lambda u(x), \qquad x \in \partial\Omega \end{cases} \tag{1.3}$$

where $c(x)$ satisfies the condition C) and proved that the eigenfunctions provide a complete orthonormal bases of certain closed subspace of $H^1(\Omega)$. Using sub and super-solutions method, Amann (see [3]), Mawhin and Schmitt (see [4]) obtained some existence results for the problem (1.1). However, since it is based on comparison techniques, the sub and super-method does not apply when the nonlinearities are compared with higher eigenvalues.

In this paper, using the Local Linking Theorem, we obtain multiple solutions for the problem (1.1), which the nonlinearites are compared with higher Neumann eigenvalue and the first Steklov eigenvalue.

2. PRELIMINARIES AND MAIN RESULTS

Let $L^p(\Omega)$ denote the Lebesgue space with the norm $|u|_p = \left(\int_\Omega |u|^p \mathrm{d}x\right)^{\frac{1}{p}}$, and $L^p(\partial\Omega, \mathrm{d}\sigma)$ with the norm $|u|_{p,\partial\Omega} = \left(\int_{\partial\Omega} |u|^p \mathrm{d}\sigma\right)^{\frac{1}{p}}$. Obviously, the space $L^2(\Omega)$ and the space $L^2(\partial\Omega, \mathrm{d}\sigma)$ inner product are denoted by

$$(u,v) = \int_\Omega uv\mathrm{d}x, (u,v)_\partial = \int_{\partial\Omega} uv\mathrm{d}\sigma$$

$H^1(\Omega)$ is a Hilbert space under the standard inner product

$$[u,v]_1 = \int_\Omega \big(u(x)v(x) + \nabla u(x)\nabla v(x)\big)\mathrm{d}x, \tag{2.1}$$

with the associated norm $\|u\|_1$. As the function $c(x)$ satisfies the condition C), we define the weighted H^1-c-inner product by

$$[u,v]_c = \int_\Omega \big(c(x)u(x)v(x) + \nabla u(x)\nabla v(x)\big)\mathrm{d}x, \tag{2.2}$$

and the associated norm $\|u\|_c$. By Corollary 3.3 in [2], we obtain that the norm $\|u\|_c$ is equivalent to the standard norm $\|u\|_1$. As the function $c(x)$ satisfies the condition C), by Equation (2.2), we can split $H^1(\Omega) = H_0^1(\Omega) \oplus_c \left[H_0^1(\Omega)\right]^\perp$ as a direct orthogonal sum.

Now, we state the Local Linking theorem introduced by [5].

Lemma 2.1 Let X is a reflexive Banach space, $X = Y \oplus V$ with $\dim V < +\infty, \Phi \in C^1(X,\mathbb{R})$ satisfies the (PS) condition, if 1) there exists a constant $\rho > 0$ such that

$$\Phi(x) \le 0, \forall x \in V \text{ with } \|x\|_V \le \rho,$$

$\Phi(x) \ge 0, \forall x \in Y$ with $\|x\|_Y \le \rho$ 2) Φ is bounded below and $\inf_X \Phi < 0$, then the functional Φ has at least two nontrivial critical points.

Proof. See Theorem 4 in [5].

For the problem (1.3), Auchmuty (see [2]) obtained that

$$\mu_1 \oint_{\partial\Omega} (\Gamma u)^2 \, d\sigma \le \int_{\Omega} \left(|\nabla u|^2 + c(x)u^2\right) dx,$$

holds for all $u \in H^1(\Omega)$, where μ_1 is the first Steklov eigenvalue for the problem (1.3). In[6], for the Neuman eigenvalue problem

$$\begin{cases} -\Delta u(x) + c(x)u = \lambda u(x), x \in \Omega, \\ \dfrac{\partial u}{\partial v} = 0, \qquad x \in \partial\Omega, \end{cases}$$

they obtain that the above problem has a sequence of real eigenvalues

$$0 < \lambda_1 \le \lambda_2 \le \cdots \le \lambda_j \le \cdots \to \infty, \text{ as } j \to \infty, \tag{2.3}$$

with finite dimensional eigenspaces.

Assume that, $f : \Omega \times \mathbb{R} \to \mathbb{R}$, $g : \overline{\Omega} \times \mathbb{R} \to \mathbb{R}$ are Carathedory functions satisfying H1) There exist $c_1 > 0$ and $2 < q_1 < 2^* = \frac{2N}{N-2}$ such that Z $|f(x,t)| \le c_1\left(1 + |t|^{q_1 - 1}\right)$ for all $t \in \mathbb{R}$ and for a.e. $x \in \Omega$;

H2) There exist $c_2 > 0$ and $2 < q_2 < 2_* = \frac{2(N-1)}{N-2}$ such that

$|g(x,t)| \le c_2\left(1+|t|^{q_2-1}\right)$ for all $t \in \mathbb{R}$ and for a.e. $x \in \partial\Omega$;

H3) There exist $\lambda, \mu > 0$ such that

$$\liminf_{|u|\to\infty} \frac{g(x,u)}{u} \le \limsup_{|u|\to\infty} \frac{g(x,u)}{u} \le \mu < \mu_1$$

and

$$\liminf_{|u|\to\infty} \frac{f(x,u)}{u} \le \limsup_{|u|\to\infty} \frac{f(x,u)}{u} \le \lambda < \lambda_1$$ uniformly for $x \in \bar{\Omega}$ with

$\lambda_1\mu + \mu_1\lambda < \mu_1\lambda_1$.

H4) There exist a integer $k \ge 1, \delta > 0$ and four constants $\alpha, \beta, a, b \in \mathbb{R}$ such that

$$\frac{1}{2}at^2 \le G(x,t) \le \frac{1}{2}bt^2 \text{ and } \frac{1}{2}\alpha t^2 \le F(x,t) \le \frac{1}{2}\beta t^2$$

uniformly for $x \in \bar{\Omega}$, $0 < |t| < \delta$, where $G(x,t) = \int_0^t g(x,s)ds$, $F(x,t) = \int_0^t f(x,s)ds$,

$\lambda_k < \alpha \le \beta < \lambda_{k+1}, \lambda_k\mu_1 < \lambda_k a + \mu_1\alpha$ and $\lambda_{k+1}b + \mu_1\beta < \lambda_{k+1}\mu_1$

Theorem 2.2 Suppose $c(x)$ satisfies C), and H1)-H4) hold, Then the problem (1.1) has at least two distinct nontrivial solutions.

3. THE PROOFS OF THEOREM 2.2

Now, we define the functional $\Phi : H^1(\Omega) \to \mathbb{R}$

$$\Phi(u) = \frac{1}{2}\int_\Omega \left(|\nabla u|^2 + c(x)u^2\right)dx - \int_\Omega F(x,u)dx - \oint_{\partial\Omega} G(x,u)d\sigma = \frac{1}{2}\|u\|_c^2 - J(u), \tag{3.1}$$

where $J(u) = \int_\Omega F(x,u)dx + \oint_{\partial\Omega} G(x,u)d\sigma.$

Since the function $f(x,u)$ satisfies H1), $g(x,u)$ satisfies H2), by the Sobolev embedding of $H^1(\Omega)$ into $L^{\frac{2N}{N-2}}(\Omega)$, the continuity of the trace operator from $H^1(\Omega)$ into $L^{\frac{2(N-1)}{N-2}}(\partial\Omega)$ and the Holder inequalitywe obtain that the functional $\Phi(u)$ is well defined. Moreover, by Lemma 2.1, and Lemma 4.2 in [7], we obtain that $\Phi \in C^1(H^1(\Omega),\square)$, and

$$\Phi'(u)v=\int_\Omega(\nabla u\nabla v+c(x)uv)dx-\int_\Omega f(x,u)vdx-\oint_{\partial\Omega}g(x,u)vd\sigma,\forall v\in H^1(\Omega).$$

(3.2)

Furthermore, the functional $J(u)$ is weakly continuous, and $J'(u)$ is compact. Let $u=v$ in (3.2) and a simple computation, we obtain that the critical point of the functional $\Phi(u)$ is the weak solution of the problem (1.1).

Lemma 3.1 (see [7]) Assume that the function $c(x)$ satisfies the condition C), H1) and H2) hold. If $\{u_m\}$ is a (PS) sequence for the functional Φ, and $\{u_m\}$ is bounded in $H^1(\Omega)$, then $\{u_m\}$ has a strongly convergence subsequence. i.e. $\Phi(u)$ satisfies the (PS) condition.

Lemma 3.2 Assume that $c(x)$ satisfies the condition C), and H1)-H3) hold, the functional $\Phi(u)$ is coercive on $H^1(\Omega)$.

Proof By H3), we obtain that there exist some constants $0<\varepsilon<\mu$, $0<\varepsilon<\lambda$, and $M_1>0$, $C>0$ such that

$$F(x,t)\le\frac{1}{2}(\lambda-\varepsilon)t^2+c,\forall|t|\ge M_1,G(x,t)\le\frac{1}{2}(\mu-\varepsilon)t^2+c,\forall|t|\ge M_1. \quad (3.3)$$

From H1), H2) and Equation (3.3), we obtain that there exist $f_0\in L^1(\Omega),g_0\in L^1(\partial\Omega)$ such that

$$F(x,u)\le\frac{1}{2}(\lambda-\varepsilon)u^2+f_0,G(x,u)\le\frac{1}{2}(\mu-\varepsilon)u^2+g_0, \quad (3.4)$$

Hence, we obtain that

$$\begin{aligned}\Phi(u)&=\frac{1}{2}\int_{\Omega}\left(|\nabla u|^2+c(x)u^2\right)dx-\int_{\Omega}F(x,u)dx-\oint_{\partial\Omega}G(x,u)d\sigma\\ &\geq\frac{1}{2}\|u\|_c^2-\frac{1}{2}(\lambda-\varepsilon)\int_{\Omega}u^2dx-\frac{1}{2}(\mu-\varepsilon)\oint_{\partial\Omega}u^2d\sigma-\int_{\Omega}f_0dx-\oint_{\partial\Omega}g_0d\sigma\end{aligned}\qquad(3.5)$$

Assume that $\|u\|_c\to\infty,$ then using the continuity of the trace operator from $H^1(\Omega)$ into $L^2(\partial\Omega)$, we obtain either $\|u\|_\partial\to\infty,$ or $\|u\|_\partial\leq M_2,$ where M_2 is a positive constant.

Case 1 As $\|u\|_\partial\leq M_2,$ by Equation (3.5), we obtain

$$\begin{aligned}\Phi(u)&\geq\frac{1}{2}\|u\|_c^2-\frac{1}{2}\left(\frac{\lambda}{\lambda_1}-\frac{\varepsilon}{\lambda_1}\right)\|u\|_c^2-\frac{1}{2}(\mu-\varepsilon)\oint_{\partial\Omega}u^2d\sigma-\int_{\Omega}f_0dx-\oint_{\partial\Omega}g_0d\sigma\\ &\geq\left(1-\frac{\lambda}{\lambda_1}+\frac{\varepsilon}{\lambda_1}\right)\|u\|_c^2-\frac{1}{2}(\mu-\varepsilon)\oint_{\partial\Omega}u^2d\sigma-\int_{\Omega}f_0dx-\oint_{\partial\Omega}g_0d\sigma.\end{aligned}$$

Hence, we obtain that $\Phi(u)$ is coercive on $H^1(\Omega)$ since $\lambda<\lambda_1$.

Case2 As $\|u\|_\partial\to\infty,$ we have

$$\begin{aligned}\Phi(u)&\geq\left(1-\frac{\lambda}{\lambda_1}+\frac{\varepsilon}{\lambda_1}\right)\|u\|_c^2-\frac{1}{2}(\mu-\varepsilon)\oint_{\partial\Omega}u^2d\sigma-\int_{\Omega}f_0dx-\oint_{\partial\Omega}g_0d\sigma\\ &\geq\left(1-\frac{\lambda}{\lambda_1}-\frac{\mu}{\mu_1}+\frac{\varepsilon}{\lambda_1}+\frac{\varepsilon}{\mu_1}\right)\|u\|_c^2-\int_{\Omega}f_0dx-\oint_{\partial\Omega}g_0d\sigma\end{aligned}$$

By H3), we obtain $\mu_1\lambda_1-\mu_1\lambda-\mu\lambda_1>0$, then $\Phi(u)$ is coercive on $H^1(\Omega)$.

Hence we obtain that the functional $\Phi(u)$ is bounded from below, and every (PS) sequence $\{u_n\}$ is bounded in $H^1(\Omega)$. From Lemma 3.1, we obtain that $\Phi(u)$ satisfies (PS) condition and is bounded from below.

The Proof of Theorem 2.2 We write $L^2(\Omega)=V\oplus V^\perp$, where

$V=E(\lambda_1)\oplus E(\lambda_2)\oplus\cdots\oplus E(\lambda_k)$ and $V^\perp=\overline{E(\lambda_{k+1})\oplus E(\lambda_{k+2})\oplus\cdots}$.

Hence, we have

$$H^1(\Omega)=\left(V\cap H^1(\Omega)\right)\oplus\left(V^\perp\cap H^1(\Omega)\right).$$

Since $\left(V\cap H^1(\Omega)\right)$ is a finite dimensional space, by [2], we obtain that for given $\delta>0$, there is a $\rho_1>0$ such that

$$u\in\left(V\cap H^1(\Omega)\right),\|u\|_c\le\rho_1\Rightarrow|u(x)|\le\delta,\text{a.e. } x\in\Omega.$$

Hence, for each $u\in\left(V\cap H^1(\Omega)\right),\|u\|_c\le\rho_1$, by H4), we have

$$\Phi(u)=\frac{1}{2}\|u\|_c^2-\int_\Omega F(x,u)\mathrm{d}x-\oint_{\partial\Omega}G(x,u)\mathrm{d}\sigma\le\frac{1}{2}\|u\|_c^2-\frac{1}{2}(\alpha-\varepsilon)\int_\Omega u^2\mathrm{d}x-\frac{1}{2}(a-\varepsilon)\oint_{\partial\Omega}u^2\mathrm{d}\sigma$$
$$\le\frac{1}{2}\left(1-\frac{a}{\mu_1}+\frac{\varepsilon}{\mu_1}\right)\|u\|_c^2-\frac{1}{2}(\alpha-\varepsilon)\int_\Omega u^2\mathrm{d}x\le\frac{1}{2}\left(1-\frac{a}{\mu_1}-\frac{\alpha}{\lambda_k}+\frac{\varepsilon}{\mu_1}+\frac{\varepsilon}{\lambda_k}\right)\|u\|_c^2$$

We have for $\varepsilon>0$ sufficiently small,

$$\frac{1}{2}\left(1-\frac{a}{\mu_1}-\frac{\alpha}{\lambda_k}+\frac{\varepsilon}{\mu_1}+\frac{\varepsilon}{\lambda_k}\right)<0.$$

Therefore, we obtain that $\Phi(u)\le 0$.

On the other hand, let $\Omega_1=\{x\in\bar{\Omega}:|u(x)|\le\delta\}$ and $\Omega_2=\{x\in\bar{\Omega}:|u(x)|>\delta\}$, then for every $u\in\left(V^\perp\cap H^1(\Omega)\right)$ with $|u(x)|\le\delta$, by H4), we obtain that

$$\int_\Omega\left(\frac{1}{2}(\beta+\varepsilon)u^2-F(x,u)\right)\mathrm{d}x\ge 0,\text{ and }\oint_{(\partial\Omega\cap\Omega_1)}\left(\frac{1}{2}(b+\varepsilon)u^2-G(x,u)\right)\mathrm{d}x\ge 0. \quad (3.6)$$

Combining Equation (3.4), Equation (3.6), H1) and H2), we have

$$\Phi(u)=\frac{1}{2}\|u\|_c^2-\int_\Omega F(x,u)\mathrm{d}x-\oint_{\partial\Omega}G(x,u)\mathrm{d}\sigma$$

$$=\frac{1}{2}\|u\|_c^2-\int_{\Omega_1} F(x,u)\mathrm{d}x-\oint_{(\partial\Omega\cap\Omega_1)}G(x,u)\mathrm{d}\sigma-\int_{\Omega_2} F(x,u)\mathrm{d}x-\oint_{(\partial\Omega\cap\Omega_2)}G(x,u)\mathrm{d}\sigma$$

$$\geq\frac{1}{2}\|u\|_c^2-\frac{(\beta+\varepsilon)}{2}\int_\Omega u^2\mathrm{d}x-\frac{(b+\varepsilon)}{2}\oint_{\partial\Omega}u^2\mathrm{d}\sigma-c_1\|u\|_c^{q_1}-c_2\|u\|_\partial^{q_2}-\int_\Omega f_0\mathrm{d}x-\oint_{\partial\Omega}g_0\mathrm{d}\sigma$$

$$\geq\frac{1}{2}\left(1-\frac{b}{\mu_1}-\frac{\varepsilon}{\mu_1}\right)\|u\|_c^2-\frac{1}{2}(\beta+\varepsilon)\int_\Omega u^2\mathrm{d}x-c_1\|u\|_c^{q_1}-c_2\|u\|_c^{q_2}-\int_\Omega f_0\mathrm{d}x-\oint_{\partial\Omega}g_0\mathrm{d}\sigma$$

$$\geq\frac{1}{2}\left(1-\frac{b}{\mu_1}-\frac{\beta}{\lambda_{k+1}}-\frac{\varepsilon}{\mu_1}-\frac{\varepsilon}{\lambda_{k+1}}\right)\|u\|_c^2-c_1\|u\|_c^{q_1}-c_2\|u\|_c^{q_2}-\int_\Omega f_0\mathrm{d}x-\oint_{\partial\Omega}g_0\mathrm{d}\sigma$$

From Equation (2.6), we have, for $\varepsilon>0$ sufficiently small

$$\frac{1}{2}\left(1-\frac{b}{\mu_1}-\frac{\beta}{\lambda_{k+1}}-\frac{\varepsilon}{\mu_1}-\frac{\varepsilon}{\lambda_{k+1}}\right)>0.$$

Since $q_1,q_2>2$, we can choose $\rho_2>0$ sufficiently small and $u\in\left(V^{\perp}\cap H^1(\Omega)\right)$, $\|u\|_c\leq\rho_2$, such that the functional $\Phi(u)\geq 0$.

By Lemma 3.2, we obtain that $\Phi(u)$ satisfies (PS) condition and is bounded from below. If $\inf_{u\in H^1(\Omega)}\Phi(u)<0,$ then by Lemma 2.1, Φ possesses two nonzero critical point. From (3.3), we obtain that there exist two nontrivial weak solutions for the problem (1.1).

4. CONCLUSION

Using Local Linking Theorem, we obtain the existence of two nontrival weak solutions for the problem (1.1) which the nonlinearites $f(x,u)$ and $g(x,u)$ are compared with higher Neumann eigenvalue and the first Steklov eigenvalue.

ACKNOWLEDGEMENTS

This paper was supported by Shanghai Natural Science Foundation Project (No. 11ZR1424500) and Shanghai Leading Academic Discipline Project (No. XTKX2012).

REFERENCES

1. M. W. Steklov, "Sur les Problemes Fondamentaux de la Physique Mathematique," Annales Scientifiques de l'École Normale Supérieure, Vol. 19, No. 1, 1902, pp. 455-490.

2. G. Auchmuty, "Steklov Eigenproblems and Representation of Solutions of Elliptic Boundary Value Problems," Numerical Functional Analysis and Optimization, Vol. 25, No. 3-4, 2005, pp. 321-348.

3. H. Amann, "Nonlinear Elliptic Equations with Nonlinear Boundary Conditions," Proceedings of the 2nd Scheveningen Conference on Differential Equations, North-Holland Mathematics Studies, Vol. 21, 1976, pp. 43-63.

4. J. Mawhin and K. Schmitt, "Upper and Lower Solutions and Semilinear Second Order Elliptic Equations with Nonlinear Boundary Conditions," Proceedings of the Royal Society of Edinburgh: Section A, Vol. 97, 1984, pp. 199-207.

5. H. Brezis and L. Nirenberg, "Remarks on Finding Critical Points," Communications on Pure and Applied Mathematics, Vol. 44, No. 8-9, 1991, pp. 939-963.

6. C. V. Pao, "Nonlinear Parabolic and Elliptic Equations," Plenum Press, New York, 1992.

7. N. Mavinga and M. N. Nkashama, "Steklov-Neumann Eigenproblems and Nonlinear Elliptic Equations with Nonlinear Boundary Conditions," Journal of Differential Equations, Vol. 248, No. 5, 2010, pp. 1212-1229.

2

UNIQUENESS OF POSITIVE SOLUTIONS TO A CLASS OF SEMILINEAR ELLIPTIC EQUATIONS

ABSTRACT

In this article, we consider the uniqueness of positive radial solutions to the Dirichlet boundary value problem

$$\begin{cases} \Delta u + f(|x|, u) + g(|x|)x \cdot \nabla u = 0, & x \in \Omega, \\ u = 0, & x \in \partial\Omega, \end{cases}$$

where Ω denotes an annulus in $\mathbb{R}^n$ $(n \geq 3)$. The uniqueness criterion is established by applying shooting method.

KEYWORDS

positive solution, semilinear elliptic equation, uniqueness

1 INTRODUCTION

This article is concerned with the positive radial solutions to a class of semilinear elliptic equations

$$\begin{aligned} &\Delta u + f(|x|, u) + g(|x|)x \cdot \nabla u = 0, \quad x \in \Omega, \\ &u = 0, \quad x \in \partial\Omega, \end{aligned} \tag{1.1}$$

where $\Omega := \{x \mid x \in \mathbb{R}^n, a < |x| < b\}$, a and b are positive real numbers, $f \in C^1((0, +\infty) \times [0, +\infty))$ and $g : [0, +\infty) \to \mathbb{R}$ is differentiable. Equation 1.1 describes stationary states for many reaction-diffusion equations. The absence of positive solutions to the elliptic equations also means that the existing solutions oscillate, which is also important information in applications.

In recent years, there is a widespread concern over the positive solutions to the Dirichlet boundary value problem (1.1) when $g(|x|) = 0$, i.e.,

$$\begin{aligned} &\Delta u + |f(|x|, u) = 0, \ u > 0 \text{ in } \Omega, \\ &u = 0, \quad x \in \partial\Omega. \end{aligned} \tag{1.2}$$

When the nonlinear term just depends on u, the uniqueness of (1.2) has been exhaustively studied (see [1–6]). In 1985, the uniqueness of (1.2) was discussed in different domains by Ni and Nussbaum [7] to the case when f depends on $|x|$ and u, $f(|x|,u) > 0$ and $f(|x|,u)$ satisfies some growth conditions. Erbe and Tang [8] presented a new uniqueness criterion using a shooting method and Sturm comparison theorem.

So far it seems that nobody considers the uniqueness to problem (1.1). Inspired by the above articles, the aim of the present article is to establish some simple criteria for the uniqueness of positive radial solutions to problem (1.1). Obviously, what we investigate in this article has a more general form than (1.2). Although due to technical reasons, when $g(|x|) = 0$ it does not hold in this article, there exist many other $g(|x|)$ which satisfy our main result.

We now conclude this introduction by outlining the rest of this article. In Section 2, we will show the existence and uniqueness of positive solutions of the initial problem

$$u'' + h(t)u' + f(t,u) = 0,$$
$$u(a) = 0,\ u'(a) = \alpha,$$

where $\alpha > 0$. Our method is the Schauder-Tikhonov fixed point theory. The existence and uniqueness of this initial problem is important to prove our main result. In Section 3, we will give the proof of our main result, i.e., show the uniqueness of positive solutions to Equation 1.1, using a shooting method and Sturm theorem.

2 PRELIMINARIES

To consider the positive radial solutions of Equation 1.1, it is reasonable to investigate the corresponding radial equation

$$u'' + \frac{n-1}{t}u' + f(t,u) + tg(t)u' = 0,$$

where $t = |x|$. For giving a proof of uniqueness of problem (1.1), let us consider the initial problem

$$\begin{aligned} &u'' + h(t)u' + f(t,u) = 0,\ t \in [a,b], \\ &u(a) = 0, \qquad u'(a) = \alpha \end{aligned} \tag{2.1}$$

where $\alpha > 0, h(t) = \frac{n-1}{t} + tg(t)$. We shall show that problem (2.1) has a unique positive solution. By a solution to problem (1.2), we mean $u \in C^2$ and $u > 0$ for all $t \in (a, b)$. First of all, we give a well-known lemma.

Lemma 2.1 *(The Schauder-Tikhonov fixed point theorem [9]). Let* $\times$ *be a Banach space and* $K \subset X$ *be a nonempty, closed, bounded and convex set. If the operator* $T : K \to X$ *continuously maps* K *into itself and* $T(K)$ *is relatively compact in* X*, then* T *has a fixed point* $x \in K$.

Theorem 2.1 *If there exist m and M, such that* $0 < m \leq u \leq M$ *for* $u \in C([a, b], (0, \infty))$ *and*

$$m \leq \int_a^t e^{-\int_a^s h(\xi)d\xi} \left(\alpha - \int_a^s e^{\int_a^\xi h(r)dr} f(\xi, u(\xi))d\xi \right) ds\, M, \quad a < t < b. \tag{2.2}$$

Then, Equation 2.1 has a unique positive solution.

Proof. We assume that

$$X = \left\{ u \in C([a, b], (0, \infty)) : \sup_{a \leq t \leq b} |u(t)| < \infty \right\},$$

endowed with the supremum norm $||u|| = \sup_{a \leq t \leq b} |u(t)|$. Let

$$K = \{u \in X : m \leq u(t) \leq M,\ t \in (a, b)\}.$$

Define the operator $T : K \rightarrow X$, by

$$(Tu)(t) = \int_a^t e^{-\int_a^s h(\xi)d\xi} \left(\alpha - \int_a^s e^{\int_a^\xi h(r)dr} f(\xi, u)d\xi \right) ds, \quad a < t < b.$$

We shall apply the Schauder-Tikhonov theorem to prove that there exists a fixed point $u(t)$, which is a positive solution of problem (2.1), for the operator T in the non-empty closed convex set K.

We shall do it by several steps as follows:

Step 1: Check that $T : K \rightarrow K$ is well defined. Obviously, by (2.2), we have

$$m \leq Tu(t) \leq M, \quad u(t) \in K,$$

thus $T : K \rightarrow K$ is well defined.

Step 2: Verify that $T : K \rightarrow K$ *is* continuous. Note that $h(t)$, $f(t, u)$ are continuous, they are integrable on $[a, b]$, there exists a constant M_1 such that

$$0 < \int_a^t e^{-\int_a^s h(\xi)d\xi} \left(\int_a^s e^{\int_a^\xi h(r)dr} d\xi \right) ds \le M_1, \quad t \in [a,b]. \tag{2.3}$$

The function $f(t,u)$ is continuous, thus for $\forall\, \varepsilon > 0$, there exists $\delta > 0$ such that for any $u(t),v(t) \in K$ with $||u\text{-}v|| \le \delta$,

$$|f(t,u) - f(t,v)| \le \frac{\varepsilon}{M_1}.$$

From this, it follows that

$$\begin{aligned}|Tu(t) - Tv(t)| &= \left| \int_a^t e^{\int_a^s h(\xi)d\xi} \left(\alpha - \int_a^s e^{\int_a^\xi h(r)dr} f(\xi, u(\xi))\, d\xi \right) ds \right. \\ &\quad \left. - \int_a^t e^{-\int_a^s h(\xi)d\xi} \left(\alpha - \int_a^s e^{\int_a^\xi h(r)dr} f(\xi, v(\xi))\, d\xi \right) ds \right| \\ &\le \int_a^t e^{-\int_a^s h(\xi)d\xi} \left(\int_a^s e^{\int_a^\xi h(r)dr} \left| f(\xi,u) - f(\xi, v(\xi)) \right| d\xi \right) ds \\ &\le \varepsilon.\end{aligned}$$

Thus, T is continuous on K.

Step 3: We check that $T(K)$ is relatively compact in X.

Since $TK \subset K$, TK is uniformly bounded. Now, verify that TK is equicontinuous. Let $u \in K$, then we have

$$(Tu)'(t) = e^{-\int_a^t h(s)ds} \left(\alpha - \int_a^t e^{\int_a^s h(r)dr} f(s,u)ds \right). \tag{2.4}$$

Similar to (2.3), there exists a constant M_2 such that

$$|(Tu)'(t)| \le M_2, \quad a < t < b.$$

Take a sequence $\{u_n\} \subset K$, by the mean value theorem, we have

$$|\,(Tu_n)(t_1) - (Tu_n)(t_2)\,| \le M_2 |t_1 - t_2|, \quad a < t < b.$$

Thus, TK is equicontinuous. Arzela-Ascoli theorem [9] implies TK is relatively compact. Now, we have verified that $T : K \to K$ satisfies all assumptions of the Schauder-Tikhonov theorem. Thus, there exists a fixed point u which is a positive solution of problem (2.1).

Now, we are in a position to prove the uniqueness of problem (2.1). The proof of the uniqueness of solution is based on the work of [10]. Suppose that u and v are two different solutions of problem (2.1), then the function

$$\omega = u' - v'$$

is a solution of Cauchy problem

$$\omega' + h(t)\omega' = \Psi(t), \quad \omega(a) = 0,$$

where $\psi = f(t,v) - f(t,u)$. It follows that

$$\omega = e^{-\int_a^t h(s)ds} \int_a^t e^{\int_a^r h(s)ds} \Psi(r)dr.$$

Hence, we have

$$|\omega(t)| \leq e^{-\int_a^t h(s)ds} \int_a^t e^{\int_a^r h(s)ds} |\Psi(r)|dr$$
$$\leq M_3 \sup_{a \leq t \leq b} |\Psi(t)|,$$

where M_3 is a constant, such that

$$0 < e^{-\int_a^t h(s)ds} \int_a^t e^{\int_a^r h(s)ds} dr \leq M_3, \quad t \in [a, b].$$

On the other hand, since the function $f(t, u)$ is Hölder continuous with respect to the second variable on $(0, +\infty)$, we obtain, for appropriate values t_0, $L > 0$,

$$
\begin{aligned}
|\Psi(t)| &\le L|u(t)-v(t)| \\
&\le L\int_a^t |u'(s)-v'(s)|ds \\
&\le L\int_a^t |\omega(s)|ds, \quad t\in(a,t_0].
\end{aligned}
$$

From this, we have $|\omega(t)| \le M_3 L\int_a^t |\omega(s)|ds$ for $t \le t_0$. It now follows from Gronwall's inequality that $\omega \equiv 0$ for $a < t \le t_0$, consequently $u' \equiv v'$ for $t \le t_0$. We find $u(t) \equiv v(t)$ for all $t \in (a,t_0]$. With the initial point t_0 replace by $\rho > t_0$, for an appropriate valueρ, the same proof can be reapplied as often as necessary to give uniqueness of any continuation of the solution whose values lie in (a, b). The proof is complete.

3 UNIQUENESS

Theorem 3.1 *Assume that $h(t)$ and $f(t,u)$ for $a < t < b$, $u(t) > 0$, satisfy inequality (2.2) and*

$(F1)\, f(t,0) \equiv 0,\ uf_u(t,u) > f(t,u) > 0,$

$(F2) f_1(t,u) \ge 0,\ h(t)v(t) + v' \ge 0,$

$(F3) h'(t) \ge 0,$

where $v(t) = \int_a^t e^{-\int_a^r h(s)ds}dr$, then problem (1.1) has at most one positive radial solution.

Example 3.1 For the equation

$$\Delta u + \frac{A}{|x|^2}x\cdot\nabla u + u^2 = 0, \quad x\in\Omega,$$

where $-n-1 \le A \le -n+1$, $\Omega := \{x \in \mathbb{R}^n | \frac{1}{2} < |x| < 1\}$, $n \ge 3$. Let $t = |x|$, then

$$h(t) = \frac{A+n-1}{t}, \quad v(t) = \int_{\frac{1}{2}}^t e^{-\int_{\frac{1}{2}}^r \frac{A+n-1}{s}ds} dr.$$

A straightforward computation yields

$$h'(t) = -\frac{A+n-1}{t^2} \geq 0$$

and

$$h(t)v(t) + v'(t) = e^{-(A+n-1)}\left(t(A+n+1) - \frac{A+n-1}{4t}\right) \geq 0, \quad t \in \left(\frac{1}{2}, 1\right).$$

Therefore, Theorem 3.1 ensures that there exists at most one positive radial solution.

Before proving our main result, we will do some preliminaries and give some useful lemmas.

Let $u(t,\alpha)$ denote the unique solution of Equation 2.1. If $\alpha > 0$, then the solution $u(t,\alpha)$ is positive for t slightly larger than a. When it vanishes in (a, b), we define $b(\alpha)$ to be the first zero of $u(t, \alpha)$. More precisely, $b(\alpha)$ is a function of α which has the property that$u(t, \alpha) > 0$ for $t \in (a, b(\alpha))$ and $u(b(\alpha), \alpha) = 0$. Let N denote the set of $\alpha > 0$ for which the solution $u(t, \alpha)$ has a finite zero $b(\alpha)$. The variation of $u(t, \alpha)$ is defined by

$$\phi(t,\alpha) = \frac{\partial u(t,\alpha)}{\partial \alpha}$$

and satisfies

$$\phi'' + h(t)\phi' + f_u(t,u)\phi = 0, \quad \phi(a,\alpha) = 0, \quad \phi(a,\alpha) = 1. \tag{3.1}$$

Let L be the linear operator given by

$$L(\phi(t,\alpha)) = \phi'' + h(t)\phi' + f_u(t,u)\phi, \quad a \leq t \leq b(\alpha). \tag{3.2}$$

By (2.4), it is easy to show that $u(t, \alpha)$ has a unique critical point $c(\alpha)$ in $(a, b(\alpha))$, and at this point, $u(t, \alpha)$ obtains a local maximum value.

Lemma 3.1 *Assume that (F2) holds, then* $\phi(t, \alpha) > 0$ *for all* $t \in (a, c(\alpha))$.

Proof. We introduce a function

$$Q(t, \alpha) = \frac{v(t)}{v'(t)} u'(t, \alpha) \geq 0, \quad a \leq t \leq c(\alpha),$$

where

$$v(t) = \int_a^t e^{-\int_a^r h(s)ds} dr$$

and accordingly

$$v'(t) = e^{-\int_a^t h(s)ds}.$$

It is easy to see that

$$v''(t) + h(t)v'(t) = 0, \quad t \in (a, c(\alpha)).$$

Differentiating $Q(t, \alpha)$ with respect to t, we get

$$Q'(t, \alpha) = u'(t, \alpha) - \frac{v(t)}{v'(t)} f(t, u(t, \alpha))$$

and

$$Q''(t, \alpha) = \left(-h(t) - \frac{v}{v'} f_u(t, u)\right) u' - \left(2 + h(t)\frac{v(t)}{v'(t)}\right) f(t, u) - \frac{v}{v'} f_t(t, u).$$

Hence, we have

$$\begin{aligned} L(Q(t, \alpha)) &= Q''(t, \alpha) + h(t)Q'(t, \alpha) + f_u(t, u)Q(t, \alpha) \\ &= -2\left(1 + h(t)\frac{v(t)}{v'(t)}\right) f(t, u) - \frac{v}{v'} f_t(t, u). \end{aligned}$$

From hypotheses $(F\,2)$, we obtain

$$L(Q(t,\alpha)) \leq 0, \quad t \in (a, c(\alpha)). \tag{3.3}$$

Since $Q(t,\alpha) > 0$ in $t \in (a,c(\alpha))$ and inequality (3.3) holds, by the Sturm comparison principle (see [2]), we see that $Q(t,\alpha)$ oscillates faster that $\phi(t,\alpha)$. Hence, $\phi(t,\alpha)$ has no zero in $t \in (a,c(\alpha))$. From $\phi(a,\alpha) = 0$ and $\phi'(a,\alpha) = 1$, it follows that $\phi(t,\alpha) > 0$ for all $t \in (a, c(\alpha))$. The proof is complete.

Remark 3.1 Lemma 3.1 was already proved in [11]. Here we give a simpler proof, directly using Sturm comparison principle.

Now, we present a lemma which has been given to the case $g(|x|) = 0$ (see [8]). To make the article as self-contained as possible, we will give a simple proof with a slight modification to [8].

Lemma 3.2 *Assume $\alpha \in N$ and $f(t,u)$ satisfies $(F\,1)$, then*

$(H\,1)$ $\phi(t,\alpha)$ vanishes at least once and at most finitely many times in $(a,b(\alpha))$,

$(H2)$ if $0 < \alpha_1 < \alpha_2$, and at least one of $u(t,\alpha_1)$ and $u(t,\alpha_2)$ has a finite zero, then they intersect in $(a,\min\{b(\alpha_1),b(\alpha_2)\})$.

Proof. We shall prove this by contradiction. Suppose to the contrary that $\phi(t,\alpha)$ does not vanish in $(a, b(\alpha))$, then $\phi(t,\alpha) > 0$, $t \in (a, b(\alpha))$. Note that $L(\phi(t,\alpha)) = 0$, so we have

$$\left(e^{\int_a^t h(s)ds}\phi'(t,\alpha)\right)' = -e^{\int_a^t h(s)ds} f_u(t,u(t,\alpha))\phi(t,\alpha). \tag{3.4}$$

Using the definition of L, we have

$$L(u(t,\alpha)) = u''(t,\alpha) + h(t)u'(t,\alpha) + f_u(t,u)u(t,\alpha).$$

Similar to (3.4), we have

$$\left(e^{\int_a^t h(s)ds}u'(t,\alpha)\right)' = e^{\int_a^t h(s)ds}L(u(t,\alpha)) - e^{\int_a^t h(s)ds}f_u(t,u(t,\alpha))u(t,\alpha). \tag{3.5}$$

Multiply both sides of (3.4) by $u(t,\ \alpha)$ and (3.5) by $\phi(t,\ \alpha)$, then subtract the resulting identities and we have

$$\left(e^{\int_a^t h(s)ds}(\phi(t,\alpha)u'(t,\alpha) - \phi'(t,\alpha)u(t,\alpha))\right)'$$
$$= e^{\int_a^t h(s)ds}\left(uf_u(t,\alpha) - f(t,u)\right)\phi(t,\alpha). \tag{3.6}$$

By $(F1)$, we have the right side of (3.6) is positive in $(a,\ b(\alpha))$. The left side of (3.6) is then a strictly increasing function of t in (3.6). We get

$$e^{\int_a^t h(s)ds}\left(\phi(t,\alpha)u'(t,\alpha) - \phi'(t,\alpha)u(t,\alpha)\right) > 0 \quad \text{at } t = b(\alpha).$$

Thus, $e^{\int_a^{b(\alpha)} h(s)ds}\phi(b(\alpha),\alpha)u'(b(\alpha),\alpha) > 0$. However, it contradicts $u'(b(\alpha),\ \alpha) < 0$ and $\phi(b(\alpha),\alpha) \geq 0$.

Since the rest of proof can be completed by the same argument as [8], we omit them.

Lemma 3.3 *If* $(F1)$ *and* $(F\,3)$ *hold, then* $\phi(b(\alpha),\ \alpha) \neq 0$.

Proof. We shall prove this by contradiction. Suppose to the contrary that $\phi(b(\alpha),\ \alpha) = 0$. Now, we may as well define $\tau(\alpha)$ to be the last zero of $\phi(t,\ \alpha)$ in $(a,\ b(\alpha))$. By Lemma 3.1, it is easy to get $c(\alpha) \leq \tau(\alpha)$, thus $u'(\tau(\alpha), \alpha) \leq 0$ and $u'(t,\ \alpha) < 0$ for all $t \in (\tau(\alpha), b(\alpha)]$. We introduce a function

$$G(t,\alpha) = u'(t,\alpha).$$

Differentiating $G(t,\ \alpha)$ with respect to t, we get

$$G'(t,\alpha) = u''(t,\alpha) = -h(t)u' - f(t,u)$$

and

$$G''(t,\alpha) = -h'(t)u'(t,\alpha) - h(t)u''(t,\alpha) - f_u(t,u)u' - f_t(t,u).$$

Hence,

$$\begin{aligned} L(G(t,\alpha)) &= G''(t,\alpha) + h(t)G'(t,\alpha) + f_u(t,u)G(t,\alpha) \\ &= -h'(t)u'(t,\alpha) - h(t)u''(t,\alpha) - f_u(t,u)u'(t,\alpha) - f_t(t,u) \\ &\quad - h^2(t)u'(t,\alpha) - h(t)f(t,u) + f_u(t,u)u'(t,\alpha) \\ &= -h'(t)u'(t,\alpha) + f_t(t,u). \end{aligned}$$

Hence, we have

$$\left(e^{\int_a^t h(s)ds}G'(t,\alpha)\right)' = e^{\int_a^t h(s)ds}L(G(t,\alpha)) - e^{\int_a^t h(s)ds}f_u(t,u(t,\alpha))G(t,\alpha). \tag{3.7}$$

Similar to the argument of Lemma 3.2, multiply both sides of (3.4) by $G(t,\alpha)$, and (3.7) by $\phi(t,\alpha)$ then we have

$$\left(e^{\int_a^t h(s)ds}(\phi(t,\alpha)G'(t,\alpha) - \phi'(t,\alpha)G(t,\alpha))\right)' = e^{\int_a^t h(s)ds}L(G(t,\alpha))\phi(t,\alpha). \tag{3.8}$$

Note that $\phi(b(\alpha),\ \alpha) = 0$, thus integrating both sides of (3.8) from $\tau(\alpha)$ to $b(\alpha)$, we obtain

$$\begin{aligned} &\left(-e^{\int_a^{b(\alpha)} h(s)ds}\phi'(b(\alpha),\alpha)G(b(\alpha),\alpha)\right) - \left(-e^{\int_a^{\tau(\alpha)} h(s)ds}\phi'(\tau(\alpha),\alpha)G(\tau(\alpha),\alpha)\right) \\ &= \int_{\tau(\alpha)}^{b(\alpha)} e^{\int_a^t h(s)ds}L(G(t,\alpha))\phi(t,\alpha)dt. \end{aligned} \tag{3.9}$$

Since $\tau(\alpha)$ to be the last zero of $\phi(t,\ \alpha)$ in $(a,\ b(\alpha))$, the behavior of $\phi(t,\ \alpha)$ in $(\tau(\alpha),\ b(\alpha))$ can be classified into two cases as follows:

(i)ϕ $(t, \alpha) > 0$ in $(\tau(\alpha), b(\alpha))$, then the left side of (3.9) is negative, but by $(F\,3)$ the right side is positive.

It is impossible.

(ii)ϕ $(t, \alpha) < 0$ in $(\tau(\alpha), b(\alpha))$, then the left side of (3.9) is positive, but by $(F\,3)$ the right side is negative.

It is also impossible. The proof is complete.

The proof of Theorem 3.1 We will prove it as a standard process. Assume that N is a nonempty set, otherwise we have nothing to prove. Let $\alpha \in N$, then $u(b(\alpha), \alpha) = 0$. It is easy to see that $u'(b(\alpha), \alpha) \leq 0$. If $u'(b(\alpha), \alpha) = 0$, then the assumption $f(t, 0) \equiv 0$ for all $t \geq 0$, and the uniqueness of solution of initial value problems for ordinary differential equations imply that $u(t, \alpha) \equiv 0$ for all $t \in [a, b(\alpha)]$, which contradicts the initial condition of $u(t, \alpha)$. Hence, we have

$$u'(b(\alpha), \alpha) < 0, \tag{3.10}$$

and the implicit function theorem implies that $b(\alpha)$ is well-defined as a function of α in N and $b(\alpha) \in C^1(N)$. Furthermore, it follows from (3.10) that N is an open set. By Lemma 3.2, we have N is an open interval (see [8]).

Differentiate both sides of the identity $u(b(\alpha), \alpha) = 0$ with respect to α, we obtain

$$u'(b(\alpha), \alpha)b'(\alpha) + \phi(b(\alpha), \alpha) = 0.$$

From above Lemma 3.3, we have $\phi(b(\alpha),\alpha) \neq 0$. Thus, $b'(\alpha) \neq 0$, $\alpha \in N$. It means that $b'(\alpha)$ does not change sign, i.e., $b(\alpha)$ is monotone. The proof is complete.

Remark 3.2 Actually, if the functions $f(|x|,u)$ and $g(|x|)$ satisfy some suitable conditions, it is not difficult to get the existence of positive radial solutions to the Dirichlet boundary value problem (1.1). We just need that for Equation 2.1, the functions $f(|x|,u)$ and $g(|x|)$ satisfy inequality (2.2) and

$$\int_a^b e^{-\int_a^s h(\xi)d\xi}\left(\alpha - \int_a^s e^{\int_a^\xi h(r)dr} f(\xi,u)d\xi\right)ds = 0.$$

However, it seems that these assumptions are too strict.

ACKNOWLEDGEMENTS

Li thanks Zhou for enthusiastic guidance and constant encouragement. The authors were very grateful to the anonymous referees for careful reading and valuable comments. This study was partially supported by the Zhejiang Innovation Project (Grant No. T200905), ZJNSF (Grant No. R6090109) and NSFC (Grant No. 10971197).

COMPETING INTERESTS

The authors declare that they have no competing interests.

AUTHORS' CONTRIBUTIONS

CL and YZ both carried out all studies in the article and approved the final version.

REFERENCES

1. Coffman CV: Uniqueness of the positive radial solution on an annulus of the Dirichlet problem for $\Delta u - u + u^3 = 0$. *J Diff Equ* 1996, 128: 379-386.

2. Kwong MK: Uniqueness of positive solutions of $\Delta u - u + u^p = 0$ in R^n. *Arch Ration Mech Anal* 1989, 105: 243-266.

3. Erbe L, Tang M: Uniqueness of positive radial solutions of $\Delta u + K(x)\gamma(u) = 0$. *Diff Integ Equ* 1998, 11: 663-678.

4. Tang M: Uniqueness of positive radial solutions for $\Delta u - u + u^p = 0$ on an annulus. *J Diff Equ* 2003, 189: 148-160.

5. Felmer P, Martínez S, Tanaka K: Uniqueness of radially symmetric positive solutions for $-\Delta u + u = u^p$ in an annulus. *J Diff Equ* 2008, 245: 1198-1209.

6. Korman P: Uniqueness and exact multiplicity of solutions for a class of Dirichlet problems. *J Diff Equ* 2008, 244: 2602-2613.

7. Ni WM, Nussbaum RD: Uniqueness and nonuniqueness for positive radial solutions of $\Delta u + f(u,r) = 0$. *Commun Pure Appl Math* 1985, 38: 67-108.

8. Erbe L, Tang M: Uniqueness of positive radial solutions of $\Delta u + f(x u) = 0$. *Diff Integ Equ* 1998, 11: 725-743.

9. Conway JB: *A Course in Function Analysis.* Springer, New York; 1990.

10. Franchi B, Lanconelli E, Serrin J: Existence and uniqueness of nonnegative solutions of quasilinear equations in R^n. *Adv Math* 1996, 118: 177-243.

11. Ma R, An Y: Uniqueness of positive solutions of a class of O.D.E. with Robin boundary conditions. *Nonlinear Anal* 2005, 63: 273-281.

3

MILD SOLUTIONS OF FRACTIONAL SEMILINEAR INTEGRO-DIFFERENTIAL EQUATIONS ON AN UNBOUNDED INTERVAL

Adel Jawahdou

Department of Mathematics, University of Carthage,
Bizerte Preparatory Engineering Institute, Jarzouna, Tunisia

ABSTRACT

In this paper, we study the existence of mild solutions for fractional semilinear integro-differential equations in an arbitrary Banach space associated with operators generating compact semigroup on the Banach space. The arguments are based on the Schauder fixed point theorem.

KEYWORDS

Semilinear Integrodifferential Equations; Mild Solutions; Schauder Fixed Point Theorem

1. INTRODUCTION

The purpose of the present paper is to present an alternative approach to the existence of solution of fractional semilinear integro-differential equations in an arbitrary Banach space X of the form

$$\begin{cases} x^{(\alpha)}(t) = A(t)x(t) + f\left(t, x(t), \int_0^t u(t,s,x(s))\mathrm{d}s\right), \\ x(0) = x_0 \in X, \end{cases} \tag{1}$$

where $t > 0, 0 < \alpha < 1$ and $A(t): D_t \subset X \to X$ generates an evolution system $U(t,s)$, satisfying:

- $U(t,s) \in B(X)$, where $B(X)$ denotes the Banach space of bounded linear operators from X into X,
- $U(t,t) = I$ (I is the identity operator in X)
- $U(t,s)U(s,z) = U(t,z)$ for $0 \le z \le s \le t < \infty$,
- the mapping $(t,s) \to U(t,s)x$ is strongly continuous in $\{(t,s): 0 \le s \le t < \infty\}$, and $f: \mathbb{R}_+ \times X \to X$ is a given function.

Differential equations of fractional order have recently proved to be valuable tools in the modeling of many phenomena in various fields of science and engineering. This equations also serve as an tool for the description of hereditary properties of various materials and processes. For details, see [1-5]. The most important problem examined up to now is that concerning the existence of solutions of considered equations. In order to solve (1), many different methods have been applied in the literature. Most of these methods use the notion of a measure of noncompactness in Banach spaces, see [6-10]. Such a method can be to apply in this work. The method we are going to use is to reduce the existence of mild solutions of fractional semilinear integro-differential equations of type (0.1) to searching a fixed points of a suitable map on the space $C(\mathbb{R}_+, X, p(t))$ tempered by an arbitrary positive real continuous function $p(t)$ defined on $\mathbb{R}_+$. In order to prove the existence of fixed points, we shall rely on the Schauder theorem. Moreover, an application to fractional differential equations is provided to illustrate the results of this work.

2. PRELIMINARY TOOLS

In what follows, X will represent a Banach space with norm $\|\cdot\|$. Denote by $C(\mathbb{R}_+,X)$ the space of continuous functions $x:\mathbb{R}_+\to X$. Now, let us assume thet $\rho=\rho(t)$ is a given function defined and continuous on the interval $\mathbb{R}_+$ with real positive values. Denote by $C(\mathbb{R}_+,\rho(t),X)=C_\rho(\mathbb{R}_+,X)$ the Banach space consisting of all functions $x=x(t)$ defined and continuous on $\mathbb{R}_+$ with values in the Banach space X such that

$$\sup\{\rho(t)\|x(t)\|:t\geq 0\}<\infty.$$

The space $C_\rho(\mathbb{R}_+,X)$ is furnished with the following standard norm

$$\|x\|_\rho=\sup\{\rho(t)\|x(t)\|:t\geq 0\}.$$

Let us recall two facts:

- The convergence in $C_\rho(\mathbb{R}_+,X)$ is the uniform convergence in the compact intervals, i.e. x_j converge to x in $C_\rho(\mathbb{R}_+,X)$ if and only if (x_j) is uniformly convergent to x on compact subsets of $\mathbb{R}_+$.
- A subset $M\subset C_\rho(\mathbb{R}_+,X)$ is relatively compact if and only if the restrictions to $[0,T]$ of all functions from X form an equicontinuous set for each $T>0$ and $M(t)$ is relatively compact in X for each $t\in\mathbb{R}_+$,where $M(t)=\{x(t):x\in X\}$, See [11].

Definition 1 A nonempty subset $\mathcal{M}\subset C_p(\mathbb{R}_+,X)$ is said to be bounded if the there is a function $r:\mathbb{R}_+\to\mathbb{R}_+$ such that $\|x(t)\|\leq r(t)$ for each $t\geq 0$ and $x\in\mathcal{M}$.

Namely, denote by $L^1_{loc}(\mathbb{R}_+)$ the space of real functions defined and Lebesgue integrable on $\mathbb{R}_+$ and equiped with the standard norm. For $x\in L^1_{loc}(\mathbb{R}_+)$ and for a fixed number $\alpha>0$ we define the Riemann-Liouville fractional integral of order α of the function $x(t)$ by putting

$$I^\alpha x(t)=\frac{1}{\Gamma(\alpha)}\int_0^t\frac{x(s)}{(t-s)^{1-\alpha}},\quad t\geq 0.$$

It may be shown that the fractional integral operator I^α transforms the space $L^1(\mathbb{R}_+)$ into itself and has some other properties (see [6-8], for example). More generally, we can consider the operator I^α on the function space $L^1_{loc}(\mathbb{R}_+)$ consisting of real functions being locally integrable over $\mathbb{R}_+$.

The following result is well known, one can see Michalski [12]

Lemma 1 For all $\alpha>0$ and $\beta>-1$.

$$\int_0^t(t-s)^{1-\alpha}s^\beta ds\leq\frac{\Gamma(\alpha)\Gamma(\beta+1)}{\Gamma(\alpha+\beta+1)}t^{\alpha+\beta}. \tag{2}$$

Our consideration is based on following Schauder fixed point theorem.

Theorem 1 [13] Let C be a closed convex subset of the Banach space X. Suppose $F:C\to C$ and F is compact (i.e., bounded sets in C are mapped into relatively compact sets). Then, F has a fixed point in C.

3. EXISTENCE OF MILD SOLUTIONS

The following hypotheses well be needed in the sequel.

- (A) $A(t)$ is a bounded linear operator on E for each $t \geq 0$ and generates a uniformly continuous evolution system $U(t,s)$ such that

  $\sup\{\|U(t,s)\| : 0 \leq s \leq t\} < \infty.$

- (C_f) (i) $(t,x,y) \in \mathbb{R}_+ \times X \times X \mapsto f(t,x,y) \in X$ satisfies the Caratheodory type conditions, i.e. $f(.,x,y)$ is measurable for $(x,y) \in X \times X$ and $f(t,.,.)$ continuous for a.e. $t \geq 0$, (ii) there exists a continuous positive function $p : \mathbb{R}_+ \to \mathbb{R}_+$ such that

  $\|f(t,x,y)\| \leq p(t)(\|x\| + \|y\|)$

  for a.e. $t \geq 0$ and all $x, y \in X$.

- (C_u) (i) $u(t,s,x) : \mathbb{R}_+ \times \mathbb{R}_+ \times X \to X$ is continuous on

  $\mathbb{R}_+ \times \mathbb{R}_+ \times X$

- (ii) $k : (t,s) \in \mathbb{R}_+ \times \mathbb{R}_+ \to \mathbb{R}_+$ being continuous such that

$\|u(t,s,x)\| \leq k(t,s)\phi(\|x\|)$ where $\phi : \mathbb{R}_+ \to (0,\infty)$ is continuous and increasing function with

$$\int_1^\infty \frac{ds}{s + \phi(s)} = \infty. \tag{3}$$

- (iii) For all positive function ψ there exist $l \in L^1_{loc}([\mathbb{R}_+])$ such that

$$\int_0^t \frac{p(s)\hat{k}(s)}{(t-s)^{1-\alpha}} \psi(s)\,ds \leq \int_0^t l(s)\psi(s)\,ds$$

.

where $\hat{k}(s) = \max\left\{1, \int_0^t k(t,s)\,ds\right\}.$

Definition 2 [14] A continuous function $x:\mathbb{R}_+\to X$ is said to be a mild solution of (0.1) if x satisfies to

$$x(t)=U(t,0)x_0$$
$$+\frac{1}{\Gamma(\alpha)}\int_0^t\frac{U(t,s)}{(t-s)^{1-\alpha}}f\left(s,x(s),\int_0^s u(s,\tau,x(\tau))\mathrm{d}\tau\right)\mathrm{d}s.$$

Our main result is given by the following theorem.

Theorem 2 If the Banach space X is separable.

Assume that the hypotheses $((A),(C_f))$ and (C_u) are satisfied. Then for each $x_0\in X$, the problem (0.1) has at least one mild solution x in $C_p(\mathbb{R}_+,X)$, for $t\geq 0$.

Proof. Consider the operator F defined by the formula

$$(Fx)(t)=U(t,0)x_0$$
$$+\frac{1}{\Gamma(\alpha)}\int_0^t\frac{U(t,s)}{(t-s)^{1-\alpha}}f\left(s,x(s),\int_0^s u(s,\tau,x(\tau))\mathrm{d}\tau\right)\mathrm{d}s, \quad (4)$$

for $t\geq 0$ and $x\in C(\mathbb{R}_+,X)$. Let

$$r(t)=G^{-1}\left(\frac{M}{\Gamma(\alpha)}\int_0^t l(s)\mathrm{d}s\right),$$

Where

$$G(t)=\int_{M\|x_0\|}^t\frac{\mathrm{d}s}{s+\phi(s)},$$

and

$$M=\sup_{s\leq t}\{U(t,s)\}.$$

The estimate (0.3) guarantee the convergence of the integral $G(t)$. In the other hand, observe that if $\beta:\mathbb{R}_+\to\mathbb{R}_+$ is nondecreasing function, then the function $\int_0^t \frac{\beta(s)}{(t-s)^{1-\alpha}} \mathrm{d}s$ is also nondecreasing on $\mathbb{R}_+$.Therefore, the function $r(t)$ is will defined and nondecreasing on $\mathbb{R}_+$. Next, put

$$\rho(t) = \frac{1}{(t^{\alpha+1}+1)(r(t)+1)}. \tag{5}$$

Obviously, the function $\rho(t)$ is continuous, positive and decreasing. In the space $C_\rho(\mathbb{R}_+, X)$ let us consider the set

$$\Omega_\rho := \{x \in C_\rho(\mathbb{R}_+, X) : \|x\| \le r(t), t \ge 0\}. \tag{6}$$

Clearly Ω_ρ is closed convex of $C_\rho(\mathbb{R}_+, X)$. Next, let $x \in \Omega_\rho$. Applying assumptions C_f (1) and C_u (2) we have

$$\begin{aligned}
\|(Fx)(t)\| &\le \|U(t,0)x_0\| + \frac{1}{\Gamma(\alpha)}\left\|\int_0^t \frac{U(t,s)}{(t-s)^{1-\alpha}} f\left(s, x(s), \int_0^s u(s,\tau,x(\tau))\mathrm{d}\tau\right)\mathrm{d}s\right\| \\
&\le \|U(t,0)x_0\| + \frac{1}{\Gamma(\alpha)}\int_0^t \frac{\|U(t,s)\|}{(t-s)^{1-\alpha}} \left\|f\left(s, x(s), \int_0^s u(s,\tau,x(\tau))\mathrm{d}\tau\right)\right\|\mathrm{d}s \\
&\le M\|x_0\| + \frac{M}{\Gamma(\alpha)}\int_0^t \frac{p(s)}{(t-s)^{1-\alpha}}\left[\|x(s)\| + \int_0^s k(s,\tau)\phi(\|x(\tau)\|)\mathrm{d}\tau\right]\mathrm{d}s \\
&\le M\|x_0\| + \frac{M}{\Gamma(\alpha)}\int_0^t \frac{p(s)}{(t-s)^{1-\alpha}}\left[r(s) + \left(\int_0^s k(s,\tau)\mathrm{d}\tau\right)\phi(r(s))\right]\mathrm{d}s \\
&\le M\|x_0\| + \int_0^t \frac{M}{\Gamma(\alpha)} \frac{p(s)\hat{k}(s)}{(t-s)^{1-\alpha}}(r(s)+\phi(r(s)))\mathrm{d}s \\
&\le M\|x_0\| + \int_0^t \frac{M}{\Gamma(\alpha)} l(s)(r(s)+\phi(r(s)))\mathrm{d}s \\
&\le M\|x_0\| + \int_0^t r'(s)\mathrm{d}s = r(t),
\end{aligned} \tag{7}$$

From the estimate (7), we deduce that F transforms Ω_ρ into itself. In what follows we show that $F:\Omega_\rho \to \Omega_\rho$ is continuous. To do this, let us fix $x \in \Omega_\rho$ and take arbitrary sequence $(x_n) \in \Omega_\rho$ such that x_n converge to x in $C_\rho(\mathbb{R}_+, X)$. Further, let us fix $\delta > 0$. Applying the properties of ρ and F we get

$$\rho(t)\|Fx_n(t)-Fx(t)\| \le \rho(t)(\|Fx_n(t)\|+\|Fx(t)\|) \le \frac{2r(t)}{(t^{\alpha+1}+1)(r(t)+1)} < \frac{2}{(t^{\alpha+1}+1)}.$$

Then, keeping in mind that $\lim_{t\to\infty}(t^{\alpha+1}+1)=\infty$, we obtain, that there exists T so big that

$$\sup\{\rho(t)\|Fx_n(t)-Fx(t)\| : t \ge T\} < \delta. \tag{8}$$

Next, for $t \le T$, denote S the operator defined by

$$Sx(t) = f\left(t, x(t), \int_0^t u(t,s,x(s))\,ds\right).$$

For $t \le T$, we have,

$$\left\|\frac{\rho(t)U(t,s)}{(t-s)^{1-\alpha}}(Sx_n(t)-Sx(t))\right\| \le 2(r(T)+\phi(r(T)))\frac{Mp(s)\hat{k}(s)}{\Gamma(\alpha)(t-s)^{1-\alpha}} \in L^1_{loc}([0,T]).$$

Next, by the Lebesgue dominated convergence theorem and (0.8) we derive that for suitable large n we have $\|Fx_n - Fx\|_\rho \le \delta.$ this fact proves that F is continuous on Ω_ρ.

Next, from (**) we see that to prove the compactness of F, we should prove that $F\Omega_\rho \subset C_\rho(\mathbb{R}_+, X)$ is equicontinuous on $[0,T]$ and $(F\Omega_\rho)$ is relatively compact for each $T>0$ and $t \ge 0$. For any $x \in \Omega_\rho$ and $t_1, t_2 \in [0,T]$ $t_2 \ge t_1$ we get,

$$\|\rho(t_2)(Fx)(t_2)-\rho(t_1)(Fx)(t_1)\| \le \|\rho(t_2)U(t_2,s)-\rho(t_1)U(t_1,s)\|\|x_0\|$$
$$+\frac{1}{\Gamma(\alpha)}\left\|\rho(t_2)\int_0^{t_2}\frac{U(t_2,s)Sx(s)}{(t_2-s)^{1-\alpha}}ds-\rho(t_1)\int_0^{t_1}\frac{U(t_1,s)Sx(s)}{(t_1-s)^{1-\alpha}}ds\right\|$$
$$\le \|\rho(t_2)U(t_2,s)-\rho(t_1)U(t_1,s)\|\|x_0\|+\frac{1}{\Gamma(\alpha)}\left\|\rho(t_2)\int_0^{t_2}\frac{U(t_2,s)Sx(s)}{(t_2-s)^{1-\alpha}}ds-\rho(t_2)\int_0^{t_2}\frac{U(t_1,s)Sx(s)}{(t_2-s)^{1-\alpha}}ds\right\|$$
$$+\frac{1}{\Gamma(\alpha)}\left\|\rho(t_2)\int_0^{t_2}\frac{U(t_1,s)Sx(s)}{(t_2-s)^{1-\alpha}}ds-\rho(t_2)\int_0^{t_1}\frac{U(t_1,s)Sx(s)}{(t_2-s)^{1-\alpha}}ds\right\|$$
$$+\frac{1}{\Gamma(\alpha)}\left\|\rho(t_2)\int_0^{t_1}\frac{U(t_1,s)Sx(s)}{(t_2-s)^{1-\alpha}}ds-\rho(t_2)\int_0^{t_1}\frac{U(t_1,s)Sx(s)}{(t_1-s)^{1-\alpha}}ds\right\|$$
$$+\frac{1}{\Gamma(\alpha)}\left\|\rho(t_2)\int_0^{t_1}\frac{U(t_1,s)Sx(s)}{(t_1-s)^{1-\alpha}}ds-\rho(t_1)\int_0^{t_1}\frac{U(t_1,s)Sx(s)}{(t_1-s)^{1-\alpha}}ds\right\|.$$

Thus,

$$\|\rho(t_2)(Fx)(t_2)-\rho(t_1)(Fx)(t_1)\| \le \|\rho(t_2)U(t_2,s)-\rho(t_1)U(t_1,s)\|\|x_0\|$$
$$+\frac{1}{\Gamma(\alpha)}\int_0^{t_2}\frac{\rho(t_2)}{(t_2-s)^{1-\alpha}}\|U(t_2,s)-U(t_1,s)\|\|Sx(s)\|ds+\frac{\rho(t_2)}{\Gamma(\alpha)}\int_{t_1}^{t_2}\frac{\|U(t_1,s)\|\|Sx(s)\|}{(t_2-s)^{1-\alpha}}ds$$
$$+\frac{\rho(t_2)}{\Gamma(\alpha)}\int_0^{t_1}\|U(t_1,s)\|\|Sx(s)\|\left[(t_2-s)^{\alpha-1}-(t_1-s)^{\alpha-1}\right]ds+\frac{1}{\Gamma(\alpha)}|\rho(t_2)-\rho(t_1)|\int_0^{t_1}\frac{\|U(t_1,s)\|\|Sx(s)\|}{(t_2-s)^{1-\alpha}}ds. \tag{9}$$

Observe that for any $\epsilon>0$ there exists $\delta>0$ such that $\|U(t_2,s)-U(t_1,s)\|\|Sx(s)\|<\epsilon$ for all $x\in\Omega_\rho$,

$s\in[0,T]$ and t_2, $t_1\in[0,T]$ such that $|t_2-t_1|\le\delta.$

Then, by the monotonicity of ρ and for all $t\ge0$ $\rho(t)\le1$, we get

$$\|\rho(t_2)(Fx)(t_2)-\rho(t_1)(Fx)(t_1)\| \le \|\rho(t_2)U(t_2,s)-\rho(t_1)U(t_1,s)\|\|x_0\|$$
$$+\frac{\epsilon B\left(r(T)+\phi(r(T))\right)}{\Gamma(\alpha)}\int_0^{t_2}\frac{ds}{(t_2-s)^{1-\alpha}}+\frac{BM\left(r(T)+\phi(r(T))\right)}{\Gamma(\alpha)}\int_{t_1}^{t_2}\frac{ds}{(t_2-s)^{1-\alpha}}$$
$$+\frac{BM\left(r(T)+\phi(r(T))\right)}{\Gamma(\alpha)}\int_0^{t_1}\left[(t_2-s)^{\alpha-1}-(t_1-s)^{\alpha-1}\right]ds$$
$$+\frac{BM\left(r(T)+\phi(r(T))\right)}{\Gamma(\alpha)}|\rho(t_2)-\rho(t_1)|\int_0^{t_1}\frac{ds}{(t_2-s)^{1-\alpha}}, \tag{10}$$

where $B=\sup_{s\in[0,T]} p(s)\hat{k}(s)$. Keeping in mind the continuity of ρ, the right-hand side of the above inequality tends to zero as $t_2 \to t_1$.

- If $t_1, t_2 \geq T$, then we have

$$\begin{aligned}
&\|\rho(t_2)(Fx)(t_2)-\rho(t_1)(Fx)(t_1)\| \\
&\leq \|\rho(t_2)(Fx)(t_2)\|+\|\rho(t_1)(Fx)(t_1)\| \\
&\leq \frac{1}{t_2^{\alpha+1}}+\frac{1}{t_1^{\alpha+1}} \leq \epsilon
\end{aligned} \tag{11}$$

- If $0<t_1<T<t_2$, note that $t_2 \to t_1$ implies that $t_2 \to T$ and $T \to t_1$. According to the above results, we have

$$\begin{aligned}
&\|\rho(t_2)(Fx)(t_2)-\rho(t_1)(Fx)(t_1)\| \\
&\leq \|\rho(t_2)(Fx)(t_2)-\rho(T)(Fx)(T)\| \\
&+\|\rho(T)(Fx)(T)-\rho(t_1)(Fx)(t_1)\|
\end{aligned} \tag{12}$$

converging to 0 as $t_1 \to t_2$.

So for $t_1, t_2 \geq 0$, $\{\rho(t)F\Omega_\rho\}$ is equicontinuous. Meanwhile, $F(\Omega_\rho)$ is relatively compact because that $F(\Omega_\rho)\subset\Omega_\rho$ is uniformly bounded. Thus F is completely continuous on Ω_ρ. By Schauder fixed point theorem, we deduce that F has a fixed point x in Ω_ρ.

The last result in this article is to prove the existence of solutions to (0.1) but with the following conditions.

- $(t,x,y)\in\mathbb{R}_+\times X\times X \mapsto f(t,x,y)\in X$ satisfies the Caratheodory type conditions, i.e. $f(.,x,y)$ is measurable for $(x,y)\in X\times X$ and $f(t,.,.)$ continuous for a.e. $t\geq 0$

$\|f(t,x,y)\| \leq t^{\frac{\alpha}{2}}$ for a.e. $t \geq 0$ and all $x, y \in X$.

Theorem 3 If the Banach space X is separable. Assume that the hypotheses A and C'_f are satisfied. Then for each $x_0 \in X$, the problem (0.1) has at least one mild solution x in $C(\mathbb{R}_+, E)$, for $t \geq 0$.

Proof. Define the operator H by:

$$(Hx)(t) = U(t,0)x_0 + \frac{1}{\Gamma(\alpha)} \int_0^t \frac{U(t,s)}{(t-s)^{1-\alpha}} f\left(s, x(s), \int_0^s u(s,\tau,x(\tau))\,d\tau\right) ds, \tag{13}$$

for $t \geq 0$. Kipping in mind the result of lemma (0.2), we get, for $x \in C(\mathbb{R}_+, E)$.

$$\begin{aligned}\|(Hx)(t)\| &\leq \|U(t,0)x_0\| + \frac{1}{\Gamma(\alpha)} \left\| \int_0^t \frac{U(t,s)}{(t-s)^{1-\alpha}} f\left(s, x(s), \int_0^s u(s,\tau,x(\tau))\,d\tau\right) ds \right\| \\ &\leq \|U(t,0)x_0\| + \frac{1}{\Gamma(\alpha)} \int_0^t \frac{\|U(t,s)\|}{(t-s)^{1-\alpha}} \left\| f\left(s, x(s), \int_0^s u(s,\tau,x(\tau))\,d\tau\right) \right\| ds \\ &\leq M\|x_0\| + \frac{M}{\Gamma(\alpha)} \int_0^t \frac{s^{-\frac{\alpha}{2}}}{(t-s)^{1-\alpha}} \leq M\|x_0\| + \frac{M\Gamma\left(1-\frac{\alpha}{2}\right)}{\Gamma\left(\frac{\alpha}{2}+1\right)} t^{\frac{\alpha}{2}}.\end{aligned} \tag{14}$$

Put $m(t) = M\|x_0\| + \dfrac{M\Gamma\left(1-\frac{\alpha}{2}\right)}{\Gamma\left(\frac{\alpha}{2}+1\right)} t^{\frac{\alpha}{2}}$ and

$$\rho'(t) = \frac{1}{(t^{\alpha+1}+1)(m(t)+1)}$$

Next, define the set

$$B_{\rho'} := \{x \in C_{\rho'}(\mathbb{R}_+, X) : \|x\| \leq m(t), t \geq 0\}. \tag{15}$$

Then, we have that H is a self-mapping of $B_{\rho'}$. We omit the proof of continuity H and $H(B_{\rho'})$ is relatively compact, because are similar to that in Theorem 2.

4. EXAMPLE

In this section, we illustrate the main result contained in Theorem 2 by the following quadratic fractional differential equation

$$\begin{cases} \dfrac{\partial^{\frac{1}{2}} x(t,\xi)}{\partial t^{\frac{1}{3}}} = \dfrac{\partial^2 x(t;\xi)}{\partial \xi^2} + \dfrac{\sqrt{|\sin(t)|}}{1+t^2} x(t,\xi) + \sqrt{|x(t;\xi)|} \displaystyle\int_0^t \dfrac{\sqrt{|\sin(t)|}\left(e^{-t}+1\right) x(s,\xi)}{\left(1+t^2+s^2\right)\left(1+|x(s,\xi)|\right)} ds, \\ x(0,\xi) = \theta, \end{cases} \quad (16)$$

for $t \geq 0$. Let $([0,1], \mathcal{A}, P)$ be a complete probability measure space. Let $X = L^2([0,1], \mathcal{A}, P)$the space of

$\mathcal{A}$-measurable maps $v(x;t)$ with

$$\|x(t)\|_2 = \left(\int_0^1 |x(t;s)| \, dP(s)\right)^{\frac{1}{2}}.$$

Consider the operator

$$A : D(A) \subseteq X \to X$$

defined by

$$D(A) = \left\{ v \in X : \frac{\partial v}{\partial x}, \frac{\partial^2 v}{\partial x^2} \in X \text{ and } x(0) = x(1) = 0 \right\}.$$

Put $A(x) = \frac{\partial^2 x}{\partial t^2}$.

Clearly A is densely defined in X and is the infinitesimal generator of a strongly continuous semigroup

$\left(U(t,s)_{\{0<s\leq t<\infty\}}\right)$ in X. Observe that the above equation is a special case of Equation (1) if we put $\alpha - \frac{1}{3}$ and

$$f\left(t,x(t,\xi),\int_0^t u(t,s,x(s,\xi))\,\mathrm{d}s\right)=\frac{\sqrt{|\sin(t)|}}{1+t^2}x(t,\xi)$$
$$+\sqrt{|x(t;\xi)|}\int_0^t\frac{\sqrt{|\sin(t)|}\left(\mathrm{e}^{-t}+1\right)x(s,\xi)}{\left(1+t^2+s^2\right)\left(1+|x(t;\xi)|\right)}\mathrm{d}s$$

Bay using the Jensen's inequality it is not difficult to see that

$$\left\|f\left(t,x(t,\xi),\int_0^t u(t,s,x(s,\xi))\,\mathrm{d}s\right)\right\|_2$$
$$\leq\frac{\sqrt{|\sin(t)|}}{1+t^2}\left[\|x(t)\|_2+\left(\mathrm{e}^{-t}\right)\sqrt{\|x(t)\|_2}\right].$$

To check conditions (H_f) and (H_u) it is enough to take

$$\begin{cases}p(t)=\sqrt{|\sin(t)|}/\left(1+t^2\right),\\ k(t,s)=\mathrm{e}^{-t}+1,\quad\text{and}\quad\max\left(1,\mathrm{e}^{-t}+1\right)=\mathrm{e}^{-t}+1\\ \phi(s)=\sqrt{s},\quad s\geq 0.\end{cases}$$

Let be a positive function ψ defined on $\mathbb{R}_+$.

$$\int_0^t\frac{m(s)(1+q(s))}{(t-s)^{1-\alpha}}\psi(s)\,\mathrm{d}s=\int_0^t\frac{2|\sin(t)|^{1-\alpha}}{t^2(t-s)^{1-\alpha}}\psi(s)\,\mathrm{d}s$$
$$=\int_0^t\frac{2|\sin(t-u)|^{1-\alpha}}{\left(1+(t-u)^2\right)u^{1-\alpha}}\psi(t-u)\,\mathrm{d}s$$
$$\leq\int_0^t\frac{2|t-u|^{1-\alpha}}{\left(1+(t-u)^2\right)u^{1-\alpha}}\psi(t-u)\,\mathrm{d}s$$
$$\leq\int_0^t\frac{2}{\left(1+(t-u)^2\right)}\psi(t-u)\,\mathrm{d}s$$
$$\leq\int_0^t\frac{2}{\left(1+s^2\right)}\psi(s)\,\mathrm{d}s$$

Thus, on the basis of Theorem 2, we conclude that Equation (4.1) has at least one mild solution in the space

$C_p(\mathbb{R}_+, L^2([0,1]))$.

ACKNOWLEDGEMENTS

I thank the referee for their invaluable advices, comments and suggestions.

REFERENCES

1. R. P. Agarwal, M. Benchohra and S. A. Hamani, "Survey on the Existence Results for Boundary Value Problems of Nonlinear Fractional Differential Equations and Inclusions," Acta Applicandae Mathematicae, Vol. 109, No. 3, 2010, pp. 973-1033. 0H

2. M. benchohra, J. Henderson and S. K. Ntouyas, "Impulsive Differential Equations and inclusions," Hindawi Publishing Corporation, New York, 2006.

3. M. Benchohra and B. A. Slimani, "Existence and Uniqueness of Solutions to Impulsive Fractional Differential Equations," Electronic Journal of Differential Equations, Vol. 10, 2009, pp. 1-11.

4. L. Byszewski, "Theorems about the Existence and Uniqueness of Solutions of a Semilinear Evolution Nonlocal Cauchy Problem," Journal of Mathematical Analysis and Applications, Vol. 162, No. 18, 1991, pp. 494-505. 1H

5. Karczewska and S. Wedrychowicz, "Existence of Mild Solutions for Semilinear Equation of Evolution," Commentationes Mathematicae Universitatis Carolinae, Vol. 37, No. 4, 1996, pp. 695-706.

6. J. Banas and K. Goebel, "Measures of Noncompactness in Banach Spaces, Lecture Notes in Pure and Applied Mathematics," Marcel Dekker, New York, 1980.

7. J. Banas, "On Existence Theorems for Differential Equations in Banach Spaces," Bulletin of the Australian Mathematical Society, Vol. 32, No. 01, 1985, pp. 73-82. 2H

8. Jawahdou, "Mild Solutions of Functional Semilinear Evolution Volterra Integrodifferential Equations on an Unbounded Interval," Nonlinear Analysis: Real World Applications, Vol. 74, No. 18, 2011, pp. 7325-7332. 3H

9. Jawahdou and A. Karoui, "Monotonic Solutions of Nonlinear Integral Equations of Fractional Order," Canadian Applied Mathematics Quarterly, Vol. 18, No. 1, 2010, pp. 41-58.

10. L. Olszowy and S. Wedrychowicz, "Mild Solutions of Semilinear Evolution Equation on an Unbounded Interval and Their Applications," Nonlinear Analysis: Real World Applications, Vol. 72, No. 3, 2010, pp. 2119-2126. 4H

11. L. V. Kantorovich and G. P. Akilov, "Functional Analysis in Normed Spaces," Pergamon Press, Oxford, 1982.

12. M. W. Michalski, "Derivatives of Noninteger Order and Their Applications. Dissertationes Mathematicae, Polska Akademia Nauk," Instytut Matematyczny, Warszawa, 1993.

13. Corduneanu, "Integral Equations and Applications," Cambridge University Press, New York, 1990.

14. M. H. M. Rashid and Y. El-Qadri, "Semilinear Fractional Integro-Differential Equations with Compact Semigroup," Nonlinear Analysis: Real World Applications, Vol. 71, No. 12, 2009, pp. 6276-6282. 5H

4

INFINITELY MANY SOLUTIONS OF STRONGLY INDEFINITE SEMILINEAR ELLIPTIC SYSTEMS

Kuan-Ju Chen

Department of Applied Science, Naval Academy, P.O. Box 90175, Zuoying, Kaohsiung 8/303, Taiwan

ABSTRACT

We proved a multiplicity result for strongly indefinite semilinear elliptic systems $-\Delta u+u=\pm 1/(1+|x|^{a})\,|v|^{p-2}v$ in $\mathbb{R}^N$, $-\Delta v+v=\pm 1/(1+|x|^{b})\,|u|^{q-2}u$ in $\mathbb{R}^N$ where a and b are positive numbers which are in the range we shall specify later.

1. INTRODUCTION

In this paper, we shall study the existence of multiple solutions of the semilinear elliptic systems

$$-\Delta u+u=\pm\frac{1}{(1+|x|)^{a}}|v|^{p-2}v \quad \text{in } \mathbb{R}^N,$$

$$-\Delta v+v=\pm\frac{1}{(1+|x|)^{b}}|u|^{q-2}u \quad \text{in } \mathbb{R}^N, \tag{1.1}$$

where a and b are positive numbers which are in the range we shall specify later. Let us consider that the exponents $p, q > 2$ are below the critical hyperbola

$$1 > \frac{1}{p} + \frac{1}{q} > \frac{N-2}{N} \quad \text{for } N \geq 3, \tag{1.2}$$

so one of p and q could be larger than $2N/(N-2)$; for that matter, the quadratic part of the energy functional

$$I^{\pm}(u,v) = \pm\int(\nabla u \cdot \nabla v + uv)dx - \frac{1}{p}\int \frac{1}{(1+|x|)^{a}}|v|^{p}dx - \frac{1}{q}\int \frac{1}{(1+|x|)^{b}}|u|^{q}dx \tag{1.3}$$

has to be redefined, and we then need fractional Sobolev spaces.

Hence the energy functional $I^{\pm}$ is strongly indefinite, and we shall use the generalized critical point theorem of Benci [1] in a version due to Heinz [2] to find critical points of $I^{\pm}$. And there is a lack of compactness due to the fact that we are working in $\mathbb{R}^N$.

In [3], Yang shows that under some assumptions on the functions f and g there exist infinitely many solutions of the semilinear elliptic systems

$$\begin{aligned} -\Delta u + u &= \pm g(x,v) \quad \text{in } \mathbb{R}^N, \\ -\Delta v + v &= \pm f(x,u) \quad \text{in } \mathbb{R}^N. \end{aligned} \tag{1.4}$$

We shall propose herein a result similar to [3] for problem (1.1).

2. ABSTRACT FRAMEWORK AND FRACTIONAL SOBOLEV SPACES

We recall some abstract results developed in [4] or [5].

We shall work with space E^s, which are obtained as the domains of fractional powers of the operator

$$-\Delta + id : H^2(\mathbb{R}^N) \cap H^1(\mathbb{R}^N) \subset L^2(\mathbb{R}^N) \to L^2(\mathbb{R}^N). \tag{2.1}$$

Namely, $E^s = D((-\Delta + id)^{s/2})$ for $0 \le s \le 2$, and the corresponding operator is denoted by $A^s : E^s \to L^2(\mathbb{R}^N)$. The spaces E^s, the usual fractional Sobolev space $H^s(\mathbb{R}^N)$, are Hilbert spaces with inner product

$$\langle u, v\rangle_{E^s} = \int A^s u A^s v dx \tag{2.2}$$

and associates norm

$$\| u \|_{E^s}^2 = \int |A^s u|^2 dx. \tag{2.3}$$

It is known that A^s is an isomorphism, and so we denote by A^{-s} the inverse of A^s.

Now let $s, t > 0$ with $s + t = 2$. We define the Hilbert space $E = E^s \times E^t$ and the bilinear form $B : E \times E \to \mathbb{R}$ by the formula

$$B((u, v), (\phi, \psi)) = \int A^s u A^t \psi + A^s \phi A^t v. \tag{2.4}$$

Using the Cauchy-Schwarz inequality, then it is easy to see that B is continuous and symmetric. Hence B induces a self-adjoint bounded linear operator $L : E \to E$ such that

$$B(z, \eta) = \langle Lz, \eta\rangle_E, \quad \text{for } z, \eta \in E. \tag{2.5}$$

Here and in what follows $\langle \cdot, \cdot \rangle_E$ denotes the inner product in E induced by $\langle \cdot, \cdot \rangle_{E^s}$ and $\langle \cdot, \cdot \rangle_{E^t}$ on the product space E in the usual way. It is easy to see that

$$Lz = L(u, v) = (A^{-s}A^t v, A^{-t}A^s u), \quad \text{for } z = (u, v) \in E. \tag{2.6}$$

We can then prove that L has two eigenvalues -1 and 1, whose corresponding eigenspaces are

$$\begin{aligned} E^- &= \{(u, -A^{-t}A^s u) : u \in E^s\}, \quad \text{for } \lambda = -1, \\ E^+ &= \{(u, A^{-t}A^s u) : u \in E^s\}, \quad \text{for } \lambda = 1, \end{aligned} \tag{2.7}$$

which give a natural splitting $E = E^+ \oplus E^-$. The spaces E^+ and E^- are orthogonal with respect to the bilinear form B, that is,

$$B(z^+, z^-) = 0, \quad \text{for } z^+ \in E^+, z^- \in E^-. \tag{2.8}$$

We can also define the quadratic form $Q : E \to \mathbb{R}$ associated to B and L as

$$Q(z) = \frac{1}{2}B(z,z) = \frac{1}{2}\langle Lz, z\rangle_E = \int A^s u A^t v \tag{2.9}$$

for all $z = (u, v) \in E$. It follows then that

$$\frac{1}{2}\| z \|_E^2 = Q(z^+) - Q(z^-), \tag{2.10}$$

where $z = z^+ + z^-$, $z^+ \in E^+$, $z^- \in E^-$. If $z = (u, v) \in E^+$, that is, $v = A^{-t}A^s u$, we have

$$Q(z) = \frac{1}{2}\| z \|_E^2 = \frac{1}{2}\| (u, A^{-t}A^s u) \|_E^2 = \| A^s u \|^2 = \| u \|_{E^s}^2. \tag{2.11}$$

Similarly

$$Q(z) = \| A^t v \|^2 = \| v \|_{E^t}^2 \tag{2.12}$$

for $z \in E^-$.

If $w(x) := 1/(1+|x|)^c$ where c is a number satisfying the condition

$$2c > 2N - \gamma(N-2s), \quad 2 < \gamma < \frac{2N}{N-2s} \tag{2.13}$$

and $m := (2N/(N-2s))/(2N/(N-2s)-\gamma)$, it follows by (2.13) that $w \in L^m(\mathbb{R}^N)$ and by Hölder inequalities that

$$\int w(x)\,|u(x)|^\gamma dx \le |w|_m\,|u|^\gamma_{2N/(N-2s)} \le c\,|w|_m\,\|u\|^\gamma_{E^s}. \tag{2.14}$$

In the sequel $|\cdot|_m$ denotes the norm in $L^m(\mathbb{R}^N)$, and we denote by $L^\gamma(w,\mathbb{R}^N)$ the weighted function spaces with the norm defined on E^s by $|u|_{w,\gamma} = (\int w(x)\,|u(x)|^\gamma)^{1/\gamma}$. According to the properties of interpolation space, we have the following embedding theorem.

Theorem 2.1. Let $s > 0$. one defines the operator $\Theta : H^s(\mathbb{R}^N) \to H^{-s}(\mathbb{R}^N)$ as follows: for $u, \phi \in H^s(\mathbb{R}^N)$,

$$\langle \Theta(u), \phi \rangle = \int w(x)\,|u|^{\gamma-2} u\phi dx. \tag{2.15}$$

Then the inclusion of $H^s(\mathbb{R}^N)$ into $L^\gamma(w,\mathbb{R}^N)$ is compact if $2 < \gamma < 2N/(N-2s)$.

Proof. Observe that, by Hölder's inequality and (2.14), we have

$$\left|\langle \Theta(u), \phi \rangle\right| \le \int \left| w(x)^{1/\gamma'}\,|u|^{\gamma-1}\,w(x)^{1/\gamma}\phi \right| \le \left(\int w(x)\,|u|^\gamma\right)^{1/\gamma'} \left(\int w(x)\,|\phi|^\gamma\right)^{1/\gamma} < \infty, \tag{2.16}$$

where $1/\gamma + 1/\gamma' = 1$; hence Θ is well defined.

Then we will claim that Θ is compact. Since $w(x) \in L^m(\mathbb{R}^N)$, for any $\varepsilon > 0$, there exists $K > 0$, such that $(\int_{|x|>K} w(x)^m)^{1/m} < \epsilon$. Now, suppose $u_n \rightharpoonup u$ weakly in $H^s(\mathbb{R}^N)$. We estimate

$$
\begin{aligned}
&\| \Theta(u_n) - \Theta(u) \|_{H^{-s}} \\
&= \sup_{\|\phi\|_{E^s} \le 1} |\langle \Theta(u_n) - \Theta(u), \phi \rangle| \\
&= \sup_{\|\phi\|_{E^s} \le 1} \left| \int w(x) \left\{ |u_n|^{\gamma-2} u_n - |u|^{\gamma-2} u \right\} \phi \right| \\
&= \sup_{\|\phi\|_{E^s} \le 1} \left| (\gamma - 1) \int w(x)\, |\theta|^{\gamma-2} (u_n - u) \phi \right|, \quad \text{where } |\theta| \le |u_n| + |u| \\
&\le C \sup_{\|\phi\|_{E^s} \le 1} \int |w(x)| \left(|u_n|^{\gamma-2} + |u|^{\gamma-2} \right) |u_n - u|\, |\phi| \\
&\le C \sup_{\|\phi\|_{E^s} \le 1} \int \left\{ |w(x)|\, |u_n|^{\gamma-2} |u_n - u|\, |\phi| + |w(x)|\, |u|^{\gamma-2} |u_n - u|\, |\phi| \right\} \\
&\le C \Bigg(\sup_{\|\phi\|_{E^s} \le 1} \int_{|x| \le K} \left\{ |w(x)|\, |u_n|^{\gamma-2} |u_n - u|\, |\phi| + |w(x)|\, |u|^{\gamma-2} |u_n - u|\, |\phi| \right\} \\
&\quad + \sup_{\|\phi\|_{E^s} \le 1} \int_{|x| > K} \left\{ |w(x)|\, |u_n|^{\gamma-2} |u_n - u|\, |\phi| + |w(x)|\, |u|^{\gamma-2} |u_n - u|\, |\phi| \right\} \Bigg),
\end{aligned}
\tag{2.17}
$$

letting

$$
m_1 = \frac{2N/(N-2s)}{2N/(N-2s) - \gamma} = m, \quad m_2 = \frac{2N/(N-2s)}{\gamma - 2}, \quad m_3 = \frac{2N}{N-2s} = m_4, \tag{2.18}
$$

we have

$$
\frac{1}{m_1} + \frac{1}{m_2} + \frac{1}{m_3} + \frac{1}{m_4} = 1, \tag{2.19}
$$

so that by H ö lder's inequality, we observe that, for any $\varepsilon > 0$, we can choose a $K > 0$ so that the integral over $(|x| > K)$ is smaller than $\varepsilon / 2$ for all n, while for this fixed K, by strong convergence of u_n to u in $L^{2N/(N-2s)}(\mathbb{R}^N)$ on any bounded region, the integral over $(|x| \le K)$ is smaller than $\varepsilon / 2$ for n large enough. We thus have proved that $\Theta(u_n) \to \Theta(u)$ strongly in $H^{-s}(\mathbb{R}^N)$; that is, the inclusion of $H^s(\mathbb{R}^N)$ into $L^\gamma(w, \mathbb{R}^N)$ is compact if $2 < \gamma < 2N/(N-2s)$.

3. MAIN THEOREM

We consider below the problem of finding multiple solutions of the semilinear elliptic systems

$$-\Delta u + u = \pm \frac{1}{(1+|x|)^a} |v|^{p-2} v \quad \text{in } \mathbb{R}^N,$$
$$-\Delta v + v = \pm \frac{1}{(1+|x|)^b} |u|^{q-2} u \quad \text{in } \mathbb{R}^N. \tag{3.1}$$

Now if we choose $s, t > 0$, $s + t = 2$, such that

$$\left(1 - \frac{1}{q}\right) \max\{p, q\} < \frac{1}{2} + \frac{s}{N},$$
$$\left(1 - \frac{1}{p}\right) \max\{p, q\} < \frac{1}{2} + \frac{t}{N}, \tag{3.2}$$

and we assume that

(H) $2 < p < 2N/(N-2t)$, $2 < q < 2N/(N-2s)$ and a and b are positive numbers such that

$$2a > 2N - p(N-2t), \quad 2b > 2N - q(N-2s). \tag{3.3}$$

We set

$$r(x) := \frac{1}{(1+|x|)^a}, \quad s(x) := \frac{1}{(1+|x|)^b} \tag{3.4}$$

and we let

$$\alpha := \frac{2N/(N-2t)}{2N/(N-2t)-p}, \quad \beta := \frac{2N/(N-2s)}{2N/(N-2s)-q} \tag{3.5}$$

so that, under assumption (H), Theorem 2.1 holds, respectively, with $w(x) := r(x)$ and $\gamma := p$, and $w(x) := s(x)$ and $\gamma := q$; that is, the inclusion of $H^s(\mathbb{R}^N)$ into $L^q(s, \mathbb{R}^N)$ and the inclusion of $H^t(\mathbb{R}^N)$ into $L^p(r, \mathbb{R}^N)$ are compact.

If $z=(u,v)\in E=E^s\times E^t$, we let

$$I^{\pm}(u,v)=\pm\int A^s u A^t v-\frac{1}{p}\int r(x)\,|v|^p\,dx-\frac{1}{q}\int s(x)\,|u|^q\,dx \tag{3.6}$$

denote the energy of z. It is well known that under assumption (H) the energy functional $I^{\pm}(u,v)$ is well defined and continuously differentiable on E, and for all $\eta=(\phi,\psi)\in E^s\times E^t$ we have

$$\pm\int A^s u A^t\psi-\int r(x)\,|v|^{p-2}\,v\psi=0, \tag{3.7}$$

$$\pm\int A^s\phi A^t v-\int s(x)\,|u|^{q-2}\,u\phi=0, \tag{3.8}$$

and it is also well known that the critical points of $I^{\pm}$ are weak solutions of problem (3.1). The main theorem is the following.

Theorem 3.1. Under assumption (H), problem (3.1) possesses infinitely many solutions $\pm(u,v)$.

Since the functional $I^{\pm}$ are strongly indefinite, a modified multiplicity critical points theorem Heinz [2] which is the generalized critical point theorem of Benci [1] will be used. For completeness, we state the result from here.

Theorem 3.2. (see [2]) Let E be a real Hilbert space, and let $I\in C^1(E,\mathbb{R})$ be a functional with the following properties:

(i) I has the form

$$I(z)=\frac{1}{2}(Lz,z)+\varphi(z)\quad\forall z\in E, \tag{3.9}$$

where L is an invertible bounded self-adjoint linear operator in E and where $\varphi\in C^1(E,\mathbb{R})$ is such that $\varphi(0)=0$ and the gradient $\nabla\varphi: E\to E$ is a compact operator;

(ii) I is even, that is $I(-z) = I(z)$ for all $z \in E$;

(iii) I satisfies the Palais-Smale condition. Furthermore, let

$$E = E^{+} \oplus E^{-} \tag{3.10}$$

be an orthogonal splitting into L-invariant subspaces E^{+}, E^{-} such that $\pm(Lz, z) \geq 0$ for all $z \in E^{\pm}$. Then,

*(a)*suppose that there is an m-dimensional linear subspace E_m of E^{+} ($m \in \mathbb{N}$) such that for the spaces $V := E^{+}, W = E^{-} \oplus E_m$ one has

(iv) $\exists \rho_0 > 0$ such that $\inf\{I(z) : z \in V, \|z\| = \rho\} > 0$ for all $\rho \in (0, \rho_0]$;

(v) $\exists c_\infty \in \mathbb{R}$ such that $I(z) \leq c_\infty$ for all $z \in W$. Then there exist at least m pairs $(z_j, -z_j)$ of critical points of I such that $0 < I(z_j) \leq c_\infty$ ($j = 1, \dots, m$);

*(b)*a similar result holds when $E_m \subset E^{-}$, and one takes $V := E^{-}$, $W = E^{+} \oplus E_m$

It is known from Section 2 that the operator L induced by the bilinear form B is an invertible bounded self-adjoint linear operator satisfying $\pm(Lz, z)_E \geq 0$ for all $z \in E^{\pm}$. We shall need some finite dimensional subspace of E. Let (e_j), $j = 1, 2, \dots$, be a complete orthogonal system in $H^s(\mathbb{R}^N)$. Let H_n denote the finite dimensional subspaces of $H^s(\mathbb{R}^N)$ generated by (e_j), $j = 1, 2, \dots, n$. Since $A^s : H^s(\mathbb{R}^N) \to L^2(\mathbb{R}^N)$ and $A^t : H^t(\mathbb{R}^N) \to L^2(\mathbb{R}^N)$ are isomorphisms, we know that $\hat{e}_j = A^{-t} A^s e_j$, $j = 1, 2, \dots$, is a complete orthogonal system in $H^t(\mathbb{R}^N)$. Let $\hat{H}_n$ denote the finite dimensional subspaces of $H^t(\mathbb{R}^N)$ generated by $(\hat{e}_j)$, $j = 1, 2, \dots, n$. For each $n \in \mathbb{N}$, we introduce the following subspaces of E^{+} and E^{-}:

$$E_n^{+} = \text{subspace of } E^{+} \text{ generated by } (e_j, \hat{e}_j), \quad j = 1, 2, \dots, n,$$

$$E_n^{-} = \text{subspace of } E^{-} \text{ generated by } (e_j, -\hat{e}_j), \quad j = 1, 2, \dots, n. \tag{3.11}$$

Lemma 3.3. The functional $I^{\pm}$ defined in (3.6) satisfies conditions (ii), (iv), and (v) of Theorem 3.2.

Proof. Condition (ii) is an immediate consequence of the definition of $I^{\pm}$. For condition (iv), by (2.11) and Theorem 2.1, for $z \in V := E^{\pm}$,

$$\begin{aligned} I^{\pm}(z) &= \pm\int A^s u A^t v dx - \frac{1}{p}\int r(x)\,|v|^p\,dx - \frac{1}{q}\int s(x)\,|u|^q\,dx \\ &\geq \frac{1}{2}\,\|z\|_E^2 - C\,\|z\|_E^p - C\,\|z\|_E^q, \end{aligned} \tag{3.12}$$

and since $p, q > 2$, we conclude that $I^{\pm}(z) > 0$ for $z \in E^{\pm}$ with $\|z\|$ small.

Next, let us prove condition (v). Let $n \in \mathbb{N}$ be fixed, let $z \in W = E_n^{\pm} \oplus E^{\mp}$, and write $z = (u, v)$ and $z = z^+ + z^-$. We have

$$\begin{aligned} I^{\pm}(z) &= \pm[Q(z^+) + Q(z^-)] - \frac{1}{p}\int r(x)\,|v|^p\,dx - \frac{1}{q}\int s(x)\,|u|^q\,dx \\ &= -\frac{1}{2}\,\|z^{\mp}\|_E^2 + \frac{1}{2}\,\|z^{\pm}\|_E^2 - \frac{1}{p}\int r(x)\,|v|^p\,dx - \frac{1}{q}\int s(x)\,|u|^q\,dx. \end{aligned} \tag{3.13}$$

Let $z^+ = (u^+, v^+) \in E^+$ and $z^- = (u^-, v^-) \in E^-$. Then we have $v^+ = A^{-t}A^s u^+$ and $v^- = -A^{-t}A^s u^-$. Furthermore, we may write $u^{\mp} = \lambda u^{\pm} + \hat{u}$, where $\hat{u}$ is orthogonal to $u^{\pm}$ in $L^2(s, \mathbb{R}^N)$. We also have $v^{\mp} = \tau v^{\pm} + \hat{v}$, where $\hat{v}$ is orthogonal to $v^{\pm}$ in $L^2(r, \mathbb{R}^N)$. It is easy to see that either λ or τ is positive. Suppose $\lambda > 0$. Then we have

$$\begin{aligned} (1+\lambda)\int s(x)\left|u^{\pm}\right|^2 dx &= \int s(x)\left[(1+\lambda)u^{\pm} + \hat{u}\right]u^{\pm} dx \\ &\leq |u|_{s,a}\left|u^{\pm}\right|_{s,a'}. \end{aligned} \tag{3.14}$$

Using the fact that the norms in $E_n^{\pm}$ are equivalent we obtain

$$\left|u^{\pm}\right|_{s,a'} \leq C\,|u|_{s,a} \tag{3.15}$$

with constant $C > 0$ independent of u. So from (3.13) and (2.11) we obtain

$$I^{\pm}(z) \le -\frac{1}{2}\|z^{\mp}\|_E^2 + \frac{1}{2}\|z^{\pm}\|_E^2 - C\left|u^{\pm}\right|_{s,\alpha}^{\alpha}$$
$$= -\frac{1}{2}\|z^{\mp}\|_E^2 + \|u^{\pm}\|_{E^s}^2 - C\left|u^{\pm}\right|_{s,\alpha}^{\alpha}. \tag{3.16}$$

The same arguments can be applied if $\tau > 0$. So the result follows from (3.16).

A sequence $\{z_n\}$ is said to be the Palais-Smale sequence for $I^{\pm}$ (PS)-sequence for short) if $|I^{\pm}(z_n)| \le C$ uniformly in n and $\nabla I^{\pm}(z_n) \xrightarrow{n} 0$ in E^*. We say that $I^{\pm}$ satisfies the Palais-Smale condition (PS)-condition for short) if every (PS)-sequence of $I^{\pm}$ is relatively compact in E.

Lemma 3.4. Under assumption (H), the functional $I^{\pm}$ satisfies the (PS)-condition.

Proof. We first prove the boundedness of (PS)-sequences of $I^{\pm}$. Let $z_n = (u_n, v_n) \in E$ be a (PS)-sequence of $I^{\pm}$ such that

$$\left|I^{\pm}(z_n)\right| = \left|\pm\int A^s u_n A^t v_n dx - \frac{1}{p}\int r(x)\,|v_n|^p\,dx - \frac{1}{q}\int s(x)\,|u_n|^q\,dx\right| \le c, \tag{3.17}$$

$$\left|\left\langle \nabla I^{\pm}(z_n), \eta\right\rangle\right| \le \epsilon_n \|\eta\|_E \text{ where } \epsilon_n = o(1) \text{ as } n \to \infty \text{ an } \eta \in E. \tag{3.18}$$

Taking $\eta = z_n$ in (3.18), it follows from (3.17), (3.18), that

$$c + \epsilon_n \|z_n\|_E \ge -\frac{1}{p}\int r(x)\,|v_n|^p\,dx - \frac{1}{q}\int s(x)\,|u_n|^q\,dx + \frac{1}{2}\int r(x)\,|v_n|^p\,dx + \frac{1}{2}\int s(x)\,|u_n|^q\,dx$$
$$= \left(\frac{1}{2} - \frac{1}{p}\right)\int r(x)\,|v_n|^p\,dx + \left(\frac{1}{2} - \frac{1}{q}\right)\int s(x)\,|u_n|^q\,dx. \tag{3.19}$$

Next, we estimate $\|u_n\|_{E^s}$ and $\|v_n\|_{E^t}$. From (3.18) with $\eta = (\phi, 0)$, we have

$$\langle \nabla I^{\pm}(z_n), \eta \rangle = \int A^s \phi A^t v_n dx - \int s(x) |u_n|^{q-2} u_n \phi dx \leq \epsilon_n \| \phi \|_{E^s} \tag{3.20}$$

for all $\phi \in E^s$. Using H ö lder's inequality and by (3.20), we obtain

$$\begin{aligned} \left| \int A^s \phi A^t v_n dx \right| &\leq \left| \int s(x) |u_n|^{q-2} u_n \phi dx \right| + \epsilon_n \| \phi \|_{E^s} \\ &\leq \int \left| s(x)^{1/q'} |u_n|^{q-1} s(x)^{1/q} \phi \right| dx + \epsilon_n \| \phi \|_{E^s} \\ &\leq \left(\int s(x) |u_n|^q \right)^{1/q'} \left(\int s(x) |\phi|^q \right)^{1/q} + \epsilon_n \| \phi \|_{E^s} \\ &\leq \left\{ C |u_n|_{s,q}^{q-1} + C \right\} \| \phi \|_{E^s} \end{aligned} \tag{3.21}$$

for all $\phi \in E^s$, which implies that

$$\| v_n \|_{E^t} \leq C |u_n|_{s,q}^{q-1} + C . \tag{3.22}$$

Similarly, we prove that

$$\| u_n \|_{E^s} \leq C |v_n|_{r,p}^{p-1} + C . \tag{3.23}$$

Adding (3.22) and (3.23) we conclude that

$$\| u_n \|_{E^s} + \| v_n \|_{E^t} \leq C \left\{ |u_n|_{s,q}^{q-1} + |v_n|_{r,p}^{p-1} + 1 \right\} . \tag{3.24}$$

Using this estimate in (3.19), we get

$$|u_n|_{s,q}^q + |v_n|_{r,p}^p \leq C \left\{ |u_n|_{s,q}^{q-1} + |v_n|_{r,p}^{p-1} \right\} + C . \tag{3.25}$$

Since $q > q-1$ and $p > p-1$, we conclude that both $|u_n|_{s,q}$ and $|v_n|_{r,p}$ are bounded, and consequently $\| u_n \|_{E^s}$ and $\| v_n \|_{E^t}$ are also bounded in terms of (3.24).

Finally, we show that $\{z_n\}$ contains a strongly convergent subsequence. It follows from $\| u_n \|_{E^s}$ and $\| v_n \|_{E^t}$ which are bounded and Theorem 2.1 that $\{z_n\}$ contains a subsequence, denoted again by $\{z_n\} = \{(u_n, v_n)\}$, such that

$$\begin{aligned} &u_n \rightharpoonup u \text{ in } E^s, \quad v_n \rightharpoonup v \text{ in } E^t, \\ &u_n \to u \text{ in } L^q(s, \mathbb{R}^N), \quad 2 < q < \frac{2N}{N-2s}, \\ &v_n \to v \text{ in } L^p(r, \mathbb{R}^N), \quad 2 < p < \frac{2N}{N-2t}. \end{aligned} \tag{3.26}$$

It follows from (3.18) that

$$\begin{aligned} &\left| \pm \int A^s \phi A^t v_n - \int s(x) |u_n|^{q-2} u_n \phi \right| \le \epsilon_n \| \phi \|_{E^s}, \quad \phi \in E^s, \\ &\left| \pm \int A^s u_n A^t \psi - \int r(x) |v_n|^{p-2} v_n \psi \right| \le \epsilon_n \| \psi \|_{E^t}, \quad \psi \in E^t. \end{aligned} \tag{3.27}$$

Therefore,

$$\begin{aligned} \| v_n - v \|_{E^t} &= \sup \frac{\left| \int A^s \phi A^t (v_n - v) \right|}{\| \phi \|_{E^s}} \\ &\le \sup \frac{\left| \int s(x) \left(|u_n|^{q-2} u_n - |u|^{q-2} u \right) \phi \right|}{\| \phi \|_{E^s}}, \end{aligned} \tag{3.28}$$

$$\begin{aligned} \| u_n - u \|_{E^s} &= \sup \frac{\left| \int A^s (u_n - u) A^t \psi \right|}{\| \psi \|_{E^t}} \\ &= \sup \frac{\left| \int r(x) \left(|v_n|^{p-2} v_n - |v|^{p-2} v \right) \psi \right|}{\| \psi \|_{E^t}}, \end{aligned} \tag{3.29}$$

and by Theorem 2.1, we conclude that $v_n \to v$ strongly in E^t and $u_n \to u$ strongly in E^s.

Proof of Theorem 3.1. Applying Lemmas 3.3 and 3.4 and Theorem 3.2, we can obtain the conclusion of Theorem 3.1.

REFERENCES

1. Benci V: On critical point theory for indefinite functionals in the presence of symmetries. *Transactions of the American Mathematical Society* 1982, 274(2):533-572.

2. Heinz H-P: Existence and gap-bifurcation of multiple solutions to certain nonlinear eigenvalue problems. *Nonlinear Analysis: Theory, Methods & Applications* 1993, 21(6):457-484.

3. Yang J: Multiple solutions of semilinear elliptic systems. *Commentationes Mathematicae Universitatis Carolinae* 1998, 39(2):257-268.

4. de Figueiredo DG, Felmer PL: On superquadratic elliptic systems. *Transactions of the American Mathematical Society* 1994, 343(1):99-116.

5. de Figueiredo DG, Yang J: Decay, symmetry and existence of solutions of semilinear elliptic systems. *Nonlinear Analysis: Theory, Methods & Applications* 1998, 33(3):211-234.

5

PSEUDO ALMOST PERIODIC WEAK SOLUTIONS OF A SEMILINEAR ELLIPTIC EQUATION

Desheng Ji1* and Chuanyi Zhang2

[1]Department of Mathematics, Harbin Institute of Technology, West Dazhi Street, Harbin, 150001, China.

[2]Department of Mathematics, Harbin Institute of Technology, West Dazhi Street, Harbin, 150001, China.

ABSTRACT

In this paper, pseudo almost periodic functions on RN, with *N* an integer larger than 1, are introduced and some basic properties of them are studied. As an application, we investigate the pseudo almost periodicity of a weak solution of the semilinear elliptic equation $-\Delta u + \sum_{j=1}^{N} c_j \partial_j u + f(x,u) = h(x)$. In addition, a pseudo almost periodic forced pendulum equation is considered as an example.

KEYWORDS

almost periodicity, pseudo almost periodicity, semilinear elliptic equation

1 INTRODUCTION

The notion of pseudo almost periodic functions on $\mathbb{R}$ was introduced by Zhang [1–3] in the early 1990s and it is a natural generalization of Bohr almost periodicity. In the past 20 years, these functions have attracted much attention and have been applied to qualitative analysis for various kinds of equations (see [4–8] and the references therein).

The notion of Bohr almost periodicity is also suitable for functions on RN, with N>1. This notion was deeply studied for numerical-valued functions in Bochner's fundamental paper [9]. In early 1970s, Sibuya [10] and Sell [11] investigated almost periodic solutions of some linear partial differential equations. Then Zaidman [12] studied almost periodic functions from RN into a Banach space. Recent work on almost periodic second order elliptic equations in RN (see, for instance, Pankov [13], Fournier*et al.* [14] and N'Guérékata [15]) shows a revival of the interest in this topic.

It is a natural idea to introduce pseudo almost periodic functions on RN and these functions may have potential use in differential equations with more than one spatial variables. However, to the best of our knowledge, there is no literature systematically treating this yet. So in this paper, we will introduce these functions and study some of their basic properties. To illustrate their potential use, we also study the pseudo almost periodicity of a weak solution of the following semilinear elliptic equation:

$$-\Delta u + \sum_{j=1}^{N} c_j \partial_j u + f(x,u) = h(x), \quad x \in \mathbb{R}^N, \tag{1.1}$$

where cj's are real constants, *f* and *h* are pseudo almost periodic functions.

In order to study (1.1), we need to consider its almost periodic component equation at first, *i.e.*,

$$-\Delta \upsilon + \sum_{j=1}^{N} c_j \partial_j \upsilon + g(x, \upsilon) = h_1(x), \quad x \in \mathbb{R}^N, \tag{1.2}$$

where g and h1 are the almost periodic components of f and h, respectively.

For the existence of bounded weak solutions of equations (1.1) and (1.2), we refer to the work of Fournier *et al.* [14]. Our main task is the following two steps.

- We firstly show the almost periodicity of a bounded weak solution υ of (1.2). This result is an improvement of [[14], Theorem 3.2] where g(x,υ)=g(υ).
- Then, we show that the bounded weak solution u of (1.1) is pseudo almost periodic by proving that u−υ is an ergodic perturbation.

The organization of this paper is as follows. In Section 2 we recall some basic facts on Bohr almost periodic functions on $\mathbb{R}^N$. Section 3 is devoted to the pseudo almost periodic functions on $\mathbb{R}^N$. In the last section, we investigate pseudo almost periodicity of a weak solution of (1.1) and consider a pseudo almost periodic forced pendulum equation as an example.

2 BOHR ALMOST PERIODIC FUNCTIONS

Let E=R or ℂ, En be the n-dimensional real or complex Euclidean space with norm $(\sum_{i=1}^{n} x_i\overline{x_i})^{1/2}$ and production $y = \sum_{i=1}^{n} x_i y_i$ for any $x, y \in \mathbb{R}^n$. Let N be an integer larger than 1. A closed ball in $\mathbb{R}^N$ (of center x0 and radius ℓ>0) is defined by $B(x_0, \ell) = \{x \in \mathbb{R}^N$ The cubes in $\mathbb{R}^N$ are defined by $K_r = \{x \in \mathbb{R}^N : |x_j| \leq r, j = 1, 2, \ldots, N\}$ We denote the Lebesgue measure of a set S⊂RN by m(S).

Let X be a Banach space with norm $\|\cdot\|$. Let $C(\mathbb{R}^N;X)$ denote the space of all continuous functions from RN to X and $BC(\mathbb{R}^N;X)$ denote the space of all bounded functions in $C(\mathbb{R}^N;X)$. Endowed with the norm

$$\|f\|_\infty = \sup_{x\in\mathbb{R}^N} \|f(x)\|,$$

$BC(\mathbb{R}^N;X)$) is a Banach space.

We say a set P⊂RN is relatively dense in RN if there exists a number ℓ>0 such that the intersection P∩B(x,ℓ) is non-empty, for all x∈RN.

Definition 2.1 ([12])

A function f∈C(R^N;X) is called almost periodic, if for every positive number ϵ, one may find a relatively dense set T(f,ϵ) in R^N, such that

$$\|f(\cdot+\tau)-f(\cdot)\|_\infty \le \epsilon, \quad \forall \tau \in T(f,\epsilon).$$

We denote by $\mathcal{AP}(\mathbb{R}^N;X)$ the space of all such almost periodic functions.

$\mathcal{AP}(\mathbb{R}^N;X)$ contains the space, $\mathrm{Trig}(\mathbb{R}^N;X)$, which consists of functions x↦$\sum_{j=1}^{n} \exp(i\xi_j\cdot x)v_j$ where $v_j \in X$, ξ_j and x∈RN, and n is any natural number (see [[13], Proposition 1.1.3]). Let x∈AP(RN;X). Since for each $t,\tau\in\mathbb{R}^N$, $|\|x(t+\tau)\| - \|x(t)\|| \le$, it is easy to see that the function $\|x(\cdot)\| \in \mathcal{AP}(\mathbb{R}^N;\mathbb{R})$..

Lemma 2.2 ([[12], Chapter 9])

(1) $\mathcal{AP}(\mathbb{R}^N;X)$ is a Banach space.

(2)For any f∈$\mathcal{AP}(\mathbb{R}^N;X)$, *the range* $\mathcal{R}_f = \{f(x) : x\in\mathbb{R}^N\}$ *is relatively compact.*

(3) If $f \in \mathcal{AP}(\mathbb{R}^N;X)$, then it is uniformly continuous over RN.

(4) $f \in C(\mathbb{R}^N;X)$ *is almost periodic if and only if it is normal; that is, the set* {f(·+s):s∈ $\mathbb{R}^N$} *is relatively compact in* $BC(\mathbb{R}^N;X)$..

(5)If f and g both are almost periodic on RN, *then for each* ϵ>0, *the set* T(f,ϵ)∩T(g,ϵ) *is relatively dense in* $\mathbb{R}^N$.

Lemma 2.3 ([[13], p.10])

(1)For $f \in C(\mathbb{R}^N;X)$ *then*

$$M(f) = \lim_{r\to\infty} \frac{1}{(2r)^N} \int_{s+K_r} f(x)\,dx, \tag{2.1}$$

where the limit exists uniformly in $s \in \mathbb{R}^N$.

(2) $Let\, f \in \mathcal{AP}(\mathbb{R}^N;\mathbb{R}).\ If\, f \geq 0\ and\ M(f) = 0,\ then\, f \equiv 0.$

The representation (2.1) has some other useful versions. For example,

$$M(f) = \lim_{r\to\infty} \frac{1}{m(B(s,r))} \int_{B(s,r)} f(x)\,dx$$

uniformly in s∈RN. For the proof we refer the reader to [16].

Let *n* be a positive integer and Ω be a subset of En. We denote by $C(\mathbb{R}^N \times \Omega;X)$ the set of all jointly continuous functions from RN×Ω to *X*. We give the following definition.

Definition 2.4 A function f∈C(RN×Ω;X) is called almost periodic in x∈RN and uniformly on compact subsets of Ω, if for every ϵ>0 and compact K⊂Ω, there exists a relatively dense set T(f,ϵ,K) in RN such that

$$\sup_{y\in K}\|f(\cdot+\tau,y)-f(\cdot,y)\|_{\infty}\leq\epsilon,\quad \forall\tau\in T(f,\epsilon,K).$$

Denote by $AP(\mathbb{R}^N\times\Omega;X)$ the set of all such functions.

Lemma 2.5 *If* $f\in C(\mathbb{R}^N\times\Omega;X)$), *then it is uniformly continuous on* RN×K *for each compact subset* $K\subset\Omega$.

Proof For ϵ>0, let *l* be the length associated with the relatively dense set *T(f,*ϵ/3,K) in RN such that $T(f,\epsilon/3,K)\cap B(x,l)\neq\emptyset$, for all x∈RN. Since *f* is uniformly continuous on B(0,l+1)×K, there is a 1>δ>0 such that when $(x',Z'),(x'',Z'')\in B(0,l+1)\times K,|x'-x''|$, and $|Z'-Z''|<\delta$, one has

$$\|f(x',Z')-f(x'',Z'')\|<\frac{\epsilon}{3}. \tag{2.2}$$

Let $(s,Z),(t,W)\in\mathbb{R}^N\times K$ be such that

$$|s-t|<\delta,\qquad |Z-W|<\delta. \tag{2.3}$$

Choosing an $\tau\in T(f,\epsilon/3,K)\cap B(-s,l)$, then one has $s+\tau\in B(0,l)$ and $t+\tau\in B(0,l+1)$. Therefore, $(s+\tau,Z)$ and $(t+\tau,W)\in B(0,l+1)\times K$ It follows from inequalities (2.2) and (2.3) that

$$\begin{aligned}\|f(t,W)-f(s,Z)\|&\leq\|f(t,W)-f(t+\tau,W)\|+\|f(t+\tau,W)-f(s+\tau,Z)\|\\&\quad+\|f(s+\tau,Z)-f(s,Z)\|<\frac{\epsilon}{3}+\frac{\epsilon}{3}+\frac{\epsilon}{3}=\epsilon.\end{aligned}$$

The proof is completed.

3 PSEUDO ALMOST PERIODIC FUNCTIONS

Denote by $PAP_0(\mathbb{R}^N;X)$ the set of ergodic perturbations φ∈ $BC(\mathbb{R}^N;X)$ for which

$$\lim_{r\to\infty}\frac{1}{m(B(0,r))}\int_{B(0,r)}\|\varphi(x)\|\,\mathrm{d}x=0. \tag{3.1}$$

It is easy to see that $\mathcal{PAP}_0(\mathbb{R}^N;X)$is a closed subspace of $\mathcal{BC}(\mathbb{R}^N;X)$and it contains the space of functions which vanish at infinite as a proper subspace.

As (2.1), (3.1) also has another version.

Lemma 3.1 *Let* $\varphi\in\mathcal{BC}(\mathbb{R}^N;X)$. *Then* $\varphi\in\mathcal{PAP}_0(\mathbb{R}^N;X)$ *if and only if*

$$\lim_{r\to\infty}\frac{1}{m(K_r)}\int_{K_r}\|\varphi(x)\|\,\mathrm{d}x=0. \tag{3.2}$$

Proof According to the volume formulas of cubes and balls in RN, we have

$$m(B(0,r))=c(N)m(K_{r/\sqrt{N}})=C(N)m(K_r),$$

where c(N) and C(N) are positive constants.

For each r>0, the inclusion $K_{r/\sqrt{N}}\subset B(0,r)\subset K_r$ holds in $\mathbb{R}^N$ with the usual Euclidean norm. It follows that

$$\int_{K_{r/\sqrt{N}}}\|\varphi(x)\|\,\mathrm{d}x\le\int_{B(0,r)}\|\varphi(x)\|\,\mathrm{d}x\le\int_{K_r}\|\varphi(x)\|\,\mathrm{d}x.$$

Dividing the inequalities by m(B(0,r)) yields

$$\begin{aligned}\frac{1}{c(N)}\frac{1}{m(K_{r/\sqrt{N}})}\int_{K_{r/\sqrt{N}}}\|\varphi(x)\|\,\mathrm{d}x&\le\frac{1}{m(B(0,r))}\int_{B(0,r)}\|\varphi(x)\|\,\mathrm{d}x\\&\le\frac{1}{C(N)}\frac{1}{m(K_r)}\int_{K_r}\|\varphi(x)\|\,\mathrm{d}x.\end{aligned}$$

Letting r→∞ in the formula above, we will get the equivalence of equations (3.1) and (3.2). □

We use (3.1) or (3.2) in different occasions for convenience. For example, (3.1) will be used in Theorem 4.6 and (3.2) in Lemma 3.6.

Definition 3.2 A closed subset *C* of RN is said to be an ergodic zero set in RN if m(C∩Kr)/m(Kr)→0 as r→∞.

Theorem 3.3 *A function* $\varphi \in \mathcal{BC}(\mathbb{R}^N;X)$ *is in* $\mathcal{PAP}_0(\mathbb{R}^N;X)$ *if and only if for each* ϵ>0, *the set* $C_\epsilon = \{x \in \mathbb{R}^N : \|\varphi(x)\| \geq \epsilon\}$ *is an ergodic zero set in* RN.

Proof If $\varphi \in \mathcal{PAP}_0(\mathbb{R}^N;X)$, we prove for each ϵ>0 the set Cϵ is an ergodic zero set in RN, *i.e.*,

$$\lim_{r\to\infty} \frac{m(C_\epsilon \cap K_r)}{m(K_r)} = 0. \tag{3.3}$$

Suppose to the contrary that there exists ϵ0>0 such that m(Cϵ0∩Kr)/m(Kr) does not converge to 0 as r→∞. Then there exists δ>0 such that for each *n*,

$$\frac{m(C_{\epsilon_0} \cap K_{r_n})}{m(K_{r_n})} \geq \delta, \quad \text{for some } r_n \geq n.$$

Thus, we have

$$\frac{1}{m(K_{r_n})}\int_{K_{r_n}} \|\varphi(x)\|\, dx \geq \frac{1}{m(K_{r_n})}\int_{K_{r_n}\cap C_{\epsilon_0}} \|\varphi(x)\|\, dx \geq \epsilon_0 \frac{m(C_{\epsilon_0} \cap K_{r_n})}{m(K_{r_n})} \geq \epsilon_0\delta,$$

which contradicts the fact that $\varphi \in \mathcal{PAP}_0(\mathbb{R}^N;X)$. So (3.3) holds.

On the other hand, suppose that (3.3) is true. That is, for any ϵ>0, there exists r0>0 such that for r>r0,

$$\frac{m(K_r \cap C_\epsilon)}{m(K_r)} < \frac{\epsilon}{\|\varphi\|_\infty + 1}.$$

Thus,

$$\frac{1}{m(K_r)}\int_{K_r}\|\varphi(x)\|\,dx = \frac{1}{m(K_r)}\int_{K_r\cap C_\epsilon}\|\varphi(x)\|\,dx + \frac{1}{m(K_r)}\int_{K_r\setminus C_\epsilon}\|\varphi(x)\|\,dx$$
$$\le \frac{m(K_r\cap C_\epsilon)}{m(K_r)}\|\varphi\|_\infty + \frac{1}{m(K_r)}\int_{K_r\setminus C_\epsilon}\epsilon\,dx$$
$$< \epsilon + \epsilon = 2\epsilon$$

for r>r0. This implies that $\varphi \in \mathcal{PAP}_0(\mathbb{R}^N;X)$.

Lemma 3.4 *The following statements hold*:

(1)A function $\varphi \in \mathcal{BC}(\mathbb{R}^N;E)$ *is in* $\mathcal{PAP}_0(\mathbb{R}^N;E)$ *if and only if* φ^2 *is.*

(2)For each positive integer n, $\Phi \in \mathcal{BC}(\mathbb{R}^N;E)^n$ *is in* $\mathcal{PAP}_0(\mathbb{R}^N;E)^n$ *if and only if the norm function* $|\Phi(\cdot)|$ *is in* $\mathcal{PAP}_0(\mathbb{R}^N;\mathbb{R})$.

Proof (1) For any ϵ>0, it is obvious that

$$C_\epsilon(\varphi) = \{x\in\mathbb{R}^N : |\varphi(x)| \ge \epsilon\} = \{x\in\mathbb{R}^N : |\varphi(x)^2| \ge \epsilon^2\} = C_{\epsilon^2}(\varphi^2).$$

Since the ϵ>0 is arbitrary, Theorem 3.3 implies the conclusion.

(2) By (1), $\Phi = (\varphi_1,\varphi_2,\dots,\varphi_n) \in \mathcal{PAP}_0(\mathbb{R}^N;E)^n$ if and only if $\varphi_i^2 \in \mathcal{PAP}_0(\mathbb{R}^N;E)$, i=1,2,...,n, the latter is equivalent to $|\Phi(\cdot)|^2 = \sum_{i=1}^n |\varphi_i(\cdot)|^2 \in \mathcal{PAP}_0(\mathbb{R}^N;\mathbb{R})$, which, again by (1), is equivalent to |Φ(·)|∈PAP0(RN;R). □

Definition 3.5 A function $f \in C(\mathbb{R}^N;X)$ is called pseudo almost periodic if f=g+φ, where $g \in \mathcal{AP}(\mathbb{R}^N;X)$ and $\varphi \in \mathcal{PAP}_0(\mathbb{R}^N;X)$. The functions g and φ are called the almost periodic component and ergodic perturbation of *f*, respectively. Denote by $\mathcal{PAP}(\mathbb{R}^N;X)$ the space of all such pseudo almost periodic functions.

The decomposition for *f* above is unique. Otherwise, let f=g˜+φ˜ where g˜ is almost periodic and φ˜ is an ergodic perturbation. Then g−g˜=φ−φ˜. If g−g˜≡0 does not hold, then by Lemma 2.3(2), M(‖g−g˜‖)>0. But since both φ and φ˜ are ergodic perturbations, M(‖φ−φ˜‖)=0. This is a contradiction.

Lemma 3.6 *If* $f \in PAP(\mathbb{R}^N; X)$ *and g is its almost periodic component, then* $g(\mathbb{R}^N) \subset \overline{f(\mathbb{R}^N)}$. *Therefore* $\|f\|_\infty \geq \|g\|_\infty \geq \inf_{x\in\mathbb{R}^N} \|g(x)\| \geq \inf_{x\in\mathbb{R}^N} \|f(x)\|$.

Proof If $g(\mathbb{R}^N) \not\subset \overline{f(\mathbb{R}^N)}$, then there exists an $x_0 \in \mathbb{R}^N$ such that $\inf_{s\in\mathbb{R}^N} \|g(x_0) - f(s)\| > \epsilon > 0$. Since g is continuous at x_0, there is a δ>0 such that |x|<δ implies that $\inf_{s\in\mathbb{R}^N} \|g(x_0 + x) - f(x_0 + s)\| > \epsilon$. By Definition 2.1, for ϵ>0 there exists $l_{\epsilon/2}>0$ such that every cube $-x_0 + s + K_{l_{\epsilon/2}}, s \in \mathbb{R}^N$ contains at least one point $\tau \in T(g, \epsilon/2)$. Therefore, if |x|<δ, then

$$\begin{aligned}\left\|\varphi(x_0 + x + \tau)\right\| &= \left\|f(x_0 + x + \tau) - g(x_0 + x + \tau)\right\| \\ &\geq \left\|f(x_0 + x + \tau) - g(x_0 + x)\right\| - \left\|g(x_0 + x) - g(x_0 + x + \tau)\right\| \\ &\geq \inf_{s\in\mathbb{R}^N} \left\|f(x_0 + s) - g(x_0 + x)\right\| - \left\|g(x_0 + \cdot) - g(x_0 + \cdot + \tau)\right\|_\infty \\ &> \epsilon/2.\end{aligned}$$

This implies that each cube $s + K_{l_{\epsilon/2}+\delta}, s \in \mathbb{R}^N$ contains a ball with radius δ on which $\|\varphi\| > \epsilon/2$.

Now, we use (3.2) to show the contradiction. For any positive integer *n*, we divide $K_{n(l_{\epsilon/2}+\delta)}$ into n^N smaller cubes with edge length $l_{\epsilon/2}+\delta$ who do not have common inner point with each other. Then, using the volume formulas $m(K_r) = (2r)^N$ and $m(B(0,r)) = [\pi^{N/2}/\Gamma(1 + N/2)] \cdot r^N$ in $\mathbb{R}^N$ where Γ(·) is the classical Gamma function, we obtain

$$\begin{aligned}\frac{1}{m(K_{n(l_{\epsilon/2}+\delta)})} \int_{K_{n(l_{\epsilon/2}+\delta)}} \left\|\varphi(x)\right\| dx &\geq \frac{1}{m(K_{n(l_{\epsilon/2}+\delta)})} \cdot n^N \cdot m\big(B(0, \delta)\big) \cdot \frac{\epsilon}{2} \\ &= \frac{n^N}{(2n(l_{\epsilon/2} + \delta))^N} \cdot \left[\pi^{N/2}/\Gamma(1 + N/2)\right] \cdot \delta^N \cdot \frac{\epsilon}{2} \\ &= \frac{[\pi^{N/2}/\Gamma(1 + N/2)] \cdot \delta^N}{2^N (l_{\epsilon/2} + \delta)^N} \frac{\epsilon}{2} > 0.\end{aligned}$$

Note that the last term in the formula above is independent of *n*. Letting n→∞ above, because $\varphi \in \mathcal{PAP}_0(\mathbb{R}^N;X)$, we get a contradiction to (3.2).

Theorem 3.7 $\mathcal{PAP}(\mathbb{R}^N;X)$ *is a Banach space.*

Proof Obviously $\mathcal{PAP}(\mathbb{R}^N;X)$ is a normed linear subspace of $\mathcal{BC}(\mathbb{R}^N;X)$, and we only need to show it is complete. Let $\{f_n\} \subset \mathcal{PAP}(\mathbb{R}^N;X)$ be a Cauchy sequence, gn and φn be the almost periodic component and ergodic perturbation of fn, respectively. By Lemma 3.6, $\{g_n\} \subset \mathcal{AP}(\mathbb{R}^N;X)$ is Cauchy too. So is $\{\varphi_n\} \subset \mathcal{PAP}_0(\mathbb{R}^N;X)$. Since $\mathcal{AP}(\mathbb{R}^N;X)$ and $\mathcal{PAP}_0(\mathbb{R}^N;X)$ are closed in $\mathcal{BC}(\mathbb{R}^N;X)$, there exist $g \in \mathcal{AP}(\mathbb{R}^N;X)$ and $\varphi \in \mathcal{PAP}_0(\mathbb{R}^N;X)$ such that $\|g_n - g\|_\infty \to 0$ and $\|\varphi_n - \varphi\|_\infty \to 0$ as $n \to \infty$. Set $f = g + \varphi$. Then $f \in \mathcal{PAP}(\mathbb{R}^N;X)$ and $\|f_n - f\|_\infty \to 0$, as $n \to \infty$.

Definition 3.8 A function $f \in C(\mathbb{R}^N \times \Omega;X)$ is called an ergodic perturbation in x∈RN and uniformly on compact subsets of Ω, if for each compact $K \subset \Omega, f \in \mathcal{BC}(\mathbb{R}^N \times K;X)$ and

$$\lim_{r\to\infty} \frac{1}{m(B(0,r))} \int_{B(0,r)} \|\varphi(x,Z)\| \, dx = 0$$

uniformly with respect to Z∈K.

Let $\mathcal{PAP}(\mathbb{R}^N \times \Omega;X)$ denote all the functions *f* of the form

$$f = g + \varphi,$$

where $g \in \mathcal{AP}(\mathbb{R}^N \times \Omega;X)$ and $\varphi \in \mathcal{PAP}_0(\mathbb{R}^N \times \Omega;X)$.

Let *K* be a subset of Ω. A function f:RN×Ω→X is said to be continuous in z∈K and uniformly in x∈RN, if for any givenz∈K and ϵ>0, there exists a δ(z,ϵ)>0, such that z′∈K and |z′−z|<δ(z,ϵ) imply that $\|f(x,z') - f(x,z)\| < \epsilon$ for all $x \in \mathbb{R}^N$.

Lemma 3.9 *Suppose the function* $f \in \mathcal{PAP}_0(\mathbb{R}^N \times \Omega; X)$ *is continuous in* z∈K *and uniformly in* x∈RN *for a compact* K⊂Ω. *Then*

$$\lim_{r\to\infty} \frac{1}{m(B(0,r))} \int_{B(0,r)} \max_{z\in K} \|f(x,z)\|\, dx = 0.$$

Proof Since *K* is compact, for arbitrary ϵ>0 one can find a finite number, say *m*, of open balls Bk with center z(k)∈K and radius δ(z(k),ϵ2), k=1,2,...,m, such that K⊂⋃mk=1Bk and

$$\sum_{k=1}^{m} \frac{1}{m(B(0,r))} \int_{B(0,r)} \left\|f\left(x, z^{(k)}\right)\right\| dx < \frac{\epsilon}{2}, \quad \forall r > r_0. \tag{3.4}$$

Since each f(·,z(k))∈PAP0(RN;X), there is a number r0>0 such that

∑k=1m1m(B(0,r))∫B(0,r)‖‖f(x,z(k))‖‖dx<ϵ2,∀r>r0. (3.5)

It follows from (3.4) and (3.5) that

$$\frac{1}{m(B(0,r))} \int_{B(0,r)} \max_{z\in K} \|f(x,z)\|\, dx \le \frac{1}{m(B(0,r))} \int_{B(0,r)} \sum_{k=1}^{m} \left\|f\left(x, z^{(k)}\right)\right\| dx + \frac{\epsilon}{2} < \epsilon$$

for r>r0. By the arbitrariness of ϵ, we get the conclusion.

4 ALMOST PERIODIC AND PSEUDO ALMOST PERIODIC WEAK SOLUTIONS

In this section, we simplify the notations for ℝ-valued function spaces in the previous sections by omitting the value space. For example, $C(\mathbb{R}^N; \mathbb{R})$ is replaced with $C(\mathbb{R}^N)$. Also some new notations will be used in this section, *i.e.*, $C^1(\mathbb{R}^N)$ is the set of all continuous differentiable functions; $C_0^\infty(\mathbb{R}^N)$ denotes the set all smooth functions with compact support; $H^1_{\text{loc}}(\mathbb{R}^N)$ denotes the local Sobolev space of functions which and their first generalized derivatives are locally square integrable; and $L^{\infty}(\mathbb{R}^N)$ $(L^{\infty}_{\text{loc}}(\mathbb{R}^N))$ is the Lebesgue measurable (locally) bounded function space.

Now, we consider the problem

$$\begin{cases} -\Delta u + \sum_{j=1}^{N} c_j \partial_j u + f(x,u) = h(x), & x \in \mathbb{R}^N, \\ u \in C(\mathbb{R}^N), \end{cases} \tag{4.1}$$

where cj's are real constants, $f \in C(\mathbb{R}^N \times \mathbb{R})$, and $h \in L^{\infty}_{\text{loc}}(\mathbb{R}^N)$.

Definition 4.1 The function $\alpha \in H^1_{\text{loc}}(\mathbb{R}^N) \cap L^{\infty}_{\text{loc}}(\mathbb{R}^N)$ is a lower solution of (4.1) if, for every $v \in C_0^{\infty}(\mathbb{R}^N)$,

$$v \geq 0 \quad \Rightarrow \quad \int_{\mathbb{R}^N} \sum_{j=1}^{N} \partial_j \alpha (\partial_j v + c_j v)\,\mathrm{d}x + \int_{\mathbb{R}^N} f(x,\alpha) v\,\mathrm{d}x - \int_{\mathbb{R}^N} hv\,\mathrm{d}x \leq 0.$$

The function $\beta \in H^1_{\text{loc}}(\mathbb{R}^N) \cap L^{\infty}_{\text{loc}}(\mathbb{R}^N)$ is an upper solution of (4.1) if, for every $v \in C_0^{\infty}(\mathbb{R}^N)$,

$$v \geq 0 \quad \Rightarrow \quad \int_{\mathbb{R}^N} \sum_{j=1}^{N} \partial_j \beta (\partial_j v + c_j v)\,\mathrm{d}x + \int_{\mathbb{R}^N} f(x,\beta) v\,\mathrm{d}x - \int_{\mathbb{R}^N} hv\,\mathrm{d}x \geq 0.$$

For the existence of a weak solution to (4.1) and its relationship with the nonlinear term *f* and forcing term *h*, we refer to the following two lemmas.

Lemma 4.2 ([[14], Theorem 2.5]) *Assume that* α *is a lower and* β *an upper solution of* (4.1) *and* $\alpha \leq \beta$. *Then problem* (4.1) *has a weak solution* $u \in C^1(\mathbb{R}^N)$ *such that* $\alpha \leq u \leq \beta$.

Lemma 4.3 ([[14], Theorem 2.7]) *Let* $h_1, h_2, \alpha, \beta \in L^{\infty}(\mathbb{R}^N)$. *Assume that there exists* $\delta > 0$ *such that*

$$\alpha(x) \leq s \leq t \leq \beta(x) \quad \Rightarrow \quad f(x,t) - f(x,s) \geq \delta(t-s). \tag{4.2}$$

If, for $i = 1,2$, $u_i \in C^1(\mathbb{R}^N)$ *is a weak solution of*

$$-\Delta u + \sum_{j=1}^{N} c_j \partial_j u + f(x,u) = h_i(x)$$

such that $\alpha \leq u_i \leq \beta$, *then*

In the remaining of this section, we will consider the almost periodicity and pseudo almost periodicity of the bounded weak solution of problem (4.1) under the assumption that *f* and *h* have the corresponding properties.

4.1 Almost periodic weak solution

To show the pseudo almost periodicity of a weak solution *u* of problem (4.1), the process consists of two parts: obtaining its almost periodic component and then the ergodic perturbation. This subsection is devoted to obtaining the almost periodic component.

Theorem 4.4 *Under the assumptions of Lemma* 4.2, *we assume further that* $\alpha, \beta \in L^\infty(\mathbb{R}^N)$ *and there exists* $\delta > 0$ *such that*

$$\operatorname{ess\,inf} \alpha \leq s \leq t \leq \operatorname{ess\,sup} \beta \quad \Rightarrow \quad f(x,t) - f(x,s) \geq \delta(t-s), \tag{4.3}$$

where essinf (esssup) *denotes the essential infimum* (*essential supremum*) *of a function. We also assume that f is almost periodic in* $x \in \mathbb{R}^N$ *and uniformly on compact subsets of* $\mathbb{R}$, *and h is almost periodic. Then the weak solution u of* (4.1) *in Lemma* 4.2 *is unique and almost periodic.*

Proof The uniqueness of *u* follows from Lemma 4.3.

From the assumptions, we know that *f* is almost periodic in $x \in \mathbb{R}^N$ and uniformly on the compact set $[\operatorname{ess\,inf} \alpha, \operatorname{ess\,sup} \beta] \subset \mathbb{R}$. By

[[12], Theorem 9.4], there exist a Banach space Y and an almost periodic function $\vec{f}:\mathbb{R}^N \to Y$ such that

$$\|f(\cdot+\tau,u)-f(\cdot,u)\|_\infty \le \|\vec{f}(\cdot+\tau)-\vec{f}(\cdot)\|_\infty, \quad \forall\tau\in\mathbb{R}^N, \forall u\in[\operatorname{ess\,inf}\alpha, \operatorname{ess\,sup}\beta].$$

For an arbitrary ϵ>0, it follows from Lemma 2.2(5) that $T(h,\delta\frac{\epsilon}{2})\cap T(\vec{f},\delta\frac{\epsilon}{2})$ is relatively dense in RN. We will show that T$T(h,\delta\frac{\epsilon}{2})\cap T(\vec{f},\delta\frac{\epsilon}{2})$, and thus u is almost periodic on RN.

Let $\tau\in T(h,\delta\frac{\epsilon}{2})\cap T(\vec{f},\delta\frac{\epsilon}{2})$. Then u(·+τ) is a weak solution of the equation

$$-\Delta u(x+\tau)+\sum_{j=1}^{N} c_j\partial_j u(x+\tau)+f(x,u(x+\tau))$$

$$=h(x+\tau)+f(x,u(x+\tau))-f(x+\tau,u(x+\tau)).$$

Note that both u and u(·+τ) satisfy the condition in (4.3). We can take the constants essinfα and esssupβ as two functions in L∞(RN), then Lemma 4.3 implies that

$$\begin{aligned}\|u-u(\cdot+\tau)\|_\infty &\le \delta^{-1}\|h(\cdot+\tau)-h(\cdot)+f(\cdot,u(\cdot+\tau))-f(\cdot+\tau,u(\cdot+\tau))\|_\infty\\ &\le \delta^{-1}\left(\|h(\cdot+\tau)-h(\cdot)\|_\infty+\|\vec{f}(\cdot)-\vec{f}(\cdot+\tau)\|_\infty\right)\\ &\le \epsilon,\end{aligned}$$

i.e., τ∈T(u,ϵ). The proof is completed. □

Remark 4.5 Theorem 4.4 improves [[14], Theorem 3.2] where the nonlinear term f(x,u)≡f(u), which benefits from the applications of the notion of almost periodic dominance f⃗ (see [[12], Chapter 9]) and Lemma 2.2(5).

4.2 Pseudo almost periodic weak solution

Now we assume that f is pseudo almost periodic in x∈RN and uniformly on compact subsets of $\mathbb{R}$ and h is also pseudo almost periodic. Let

$$f(x,u) = g(x,u) + \varphi(x,u), \qquad h(x) = h_1(x) + h_2(x), \quad \forall x \in \mathbb{R}^N, \forall u \in \mathbb{R},$$

where g and h1 are the almost periodic components, φ and h2 are ergodic perturbations.

In the following, we shall use the auxiliary nonnegative function ψR∈C∞0(RN), R≥1, such that for some κ∈(0,1),

$$\psi_R(x) = \begin{cases} 1, & |x| \le R, \\ 0, & |x| \ge R + R^\kappa, \end{cases}$$

$$\left|\partial^\gamma \psi_R(x)\right| \le C_\gamma \cdot R^{-\kappa|\gamma|},$$

where γ=(γ1,γ2,...,γn) is the multiple index with the γi's being nonnegative integers, $\gamma| = \sum_{i=1}^N \gamma_i$, $\partial^\gamma = \frac{\partial^{|\gamma|}}{\partial x_1^{\gamma_1} \cdots \partial x_N^{\gamma_N}}$, $C_\gamma > 0$ does not depend on R. The function ψR originates from [[17], Lemma 4.1] and has been used in [[13], page 159] for another purpose.

To consider (4.1), we also need to consider its almost periodic component, *i.e.* the equation

$$\begin{cases} -\Delta v + \sum_{j=1}^N c_j \partial_j v + g(x,v) = h_1(x), & x \in \mathbb{R}^N, \\ v \in C(\mathbb{R}^N). \end{cases} \tag{4.4}$$

Theorem 4.6 *Assume that* $\alpha_1, \alpha_2 \in H^1_{loc}(\mathbb{R}^N) \cap L^\infty(\mathbb{R}^N)$ *are lower and* $\beta_1, \beta_2 \in H^1_{loc}(\mathbb{R}^N) \cap$ *upper solutions of*(4.1) *and* (4.4), *respectively. Let f be continuous in* z∈K *and uniformly in* x∈RN *for each compact* K⊂R. *If there exists* δ>0 *such that*

$$\operatorname{ess\,inf}\min\{\alpha_1,\alpha_2\} \le s \le t \le \operatorname{ess\,sup}\max\{\beta_1,\beta_2\}$$

$$\Rightarrow \quad f(x,t) - f(x,s) \ge \delta(t-s), \qquad (4.5)$$

then problem (4.1) *has a unique pseudo almost periodic weak solution* $u \in C^1(\mathbb{R}^N)$ *such that* $\alpha_1 \le u \le \beta_1$ *whose almost periodic component is a weak solution of* (4.4).

Proof It follows from Lemma 3.6 that (4.5) is also valid when *f* is replaced with *g*. In fact, for each pair of *s* and *t* satisfying the condition in (4.5), $g(\cdot,t)-g(\cdot,s)$ is the almost periodic component of the pseudo almost periodic function $f(\cdot,t)-f(\cdot,s)$. By Lemma 3.6, for each $x \in \mathbb{R}^N$ there exists a sequence $\{x_n\} \subset \mathbb{R}^N$ such that

$$g(x,t) - g(x,s) = \lim_{n\to\infty} f(x_n,t) - f(x_n,s) \ge \delta(t-s).$$

Lemmas 4.2 and 4.3 imply the existence and uniqueness of weak solutions $u \in C^1(\mathbb{R}^N)$ and $\upsilon \in C^1(\mathbb{R}^N)$ of (4.1) and (4.4), respectively, such that $\alpha_1 \le u \le \beta_1$ and $\alpha_2 \le \upsilon \le \beta_2$. Moreover, Theorem 4.4 implies that υ is almost periodic.

Now if we can show that $u-\upsilon \in PAP_0(\mathbb{R}^N)$, then the proof is completed.

Let $w=u-\upsilon$, then *w* is a weak solution of the equation

$$-\Delta w + \sum_{j=1}^{N} c_j \partial_j w + f(x,u) - f(x,\upsilon) = h_2(x) - \varphi(x,\upsilon), \quad x \in \mathbb{R}^N.$$

Since $w \in C^1(\mathbb{R}^N)$ and $\psi_R \in C_0^\infty(\mathbb{R}^N)$, $w \cdot \psi_R \in H_0^1(\mathbb{R}^N)$. Thus, by the definition of a weak solution, we have

$$\int_{B(0,R+R^{\kappa})} \nabla w \cdot \nabla(w \cdot \psi_R)\,\mathrm{d}x + \int_{B(0,R+R^{\kappa})} \sum_{j=1}^{N} c_j \partial_j w \cdot (w \cdot \psi_R)\,\mathrm{d}x$$
$$+ \int_{B(0,R+R^{\kappa})} [f(x,u) - f(x,v)](w \cdot \psi_R)\,\mathrm{d}x$$
$$= \int_{B(0,R+R^{\kappa})} [h_2(x) - \varphi(x,v)](w \cdot \psi_R)\,\mathrm{d}x. \tag{4.6}$$

Now we will give some estimations for terms on the left side of (4.6). For the first term, we have

$$I_1 = \int_{B(0,R+R^{\kappa})} \nabla w \cdot \nabla(w \cdot \psi_R)\,\mathrm{d}x$$
$$= \int_{B(0,R+R^{\kappa})} \nabla w \cdot \nabla w \cdot \psi_R\,\mathrm{d}x + \int_{B(0,R+R^{\kappa})} \nabla w \cdot w \cdot \nabla \psi_R\,\mathrm{d}x. \tag{4.7}$$

Since ψR is nonnegative, ∫B(0,R+Rκ)∇w·ψRdx≥0. Using integration by parts, it is easy to see that

$$\int_{B(0,R+R^{\kappa})} \nabla w \cdot w \cdot \nabla \psi_R\,\mathrm{d}x$$
$$= \sum_{j=1}^{N} \int_{B(0,R+R^{\kappa})} \partial_j\left(\frac{1}{2}w^2\right) \cdot \partial_j \psi_R\,\mathrm{d}x$$
$$\geq -\sum_{j=1}^{N} \left[\int_{|x-x_j e_j| \leq R+R^{\kappa}} \mathrm{d}x_1 \cdots \widehat{\mathrm{d}x_j} \cdots \mathrm{d}x_N \left\| w^2 \cdot \partial_j \psi_R \right\|_{\infty}\right.$$
$$\left. + \int_{B(0,R+R^{\kappa})} \frac{1}{2} w^2 \cdot \partial_j^2 \psi_R\,\mathrm{d}x\right]$$
$$\geq -N \cdot \left[\pi^{(N-1)/2}/\Gamma(1 + (N-1)/2)\right](R + R^{\kappa})^{N-1} \cdot \|w\|_{\infty}^2 \cdot \max_{|\gamma|=1}\{C_\gamma\} R^{-\kappa}$$
$$- \frac{1}{2} N \cdot \left[\pi^{N/2}/\Gamma(1 + N/2)\right]\left[(R + R^{\kappa})^{N} - R^{N}\right] \cdot \|w\|_{\infty}^2 \cdot \max_{|\gamma|=2}\{C_\gamma\} R^{-2\kappa}$$
$$= -\mathcal{O}(R^{N-1-\kappa}) \quad \text{as } R \to \infty,$$

where ej∈RN whose j th element is 1 and the others are 0, $\mathcal{O}$ is the symbol of infinite of the same order.

For the second term in (4.6), similar to the last formulas, we have

$$\begin{aligned}
I_2 &= \int_{B(0,R+R^{\kappa})} \sum_{j=1}^{N} c_j \partial_j w \cdot (w \cdot \psi_R)\,\mathrm{d}x = \sum_{j=1}^{N} c_j \int_{B(0,R+R^{\kappa})} \partial_j\left(\frac{1}{2}w^2\right) \cdot \psi_R\,\mathrm{d}x \\
&\geq -\sum_{j=1}^{N} |c_j| \cdot \left[\pi^{(N-1)/2}/\Gamma(1+(N-1)/2)\right](R+R^{\kappa})^{N-1} \cdot \|w\|_{\infty}^2 \cdot C_{\theta} \\
&\quad -\frac{1}{2}\sum_{j=1}^{N} |c_j| \cdot \left[\pi^{(N-1)/2}/\Gamma(1+(N-1)/2)\right]\left[(R+R^{\kappa})^{N} - R^{N}\right] \cdot \|w\|_{\infty}^2 \max_{|\gamma|=1}\{C_{\gamma}\}R^{-\kappa} \\
&= -\mathcal{O}(R^{N-1}) \quad \text{as } R \to \infty,
\end{aligned}$$

where θ=(0,...,0) is the zero index.

For the third term, it is implied by (4.5) that

$$I_3 = \int_{B(0,R+R^{\kappa})} [f(x,u) - f(x,v)](w \cdot \psi_R)\,\mathrm{d}x \geq \delta \int_{B(0,R)} w^2\,\mathrm{d}x.$$

Multiplying (4.6) by $\frac{1}{m(B(0,R+R^{\kappa}))}$ and using the estimations for I1-I3, we obtain for sufficiently large R,

$$\begin{aligned}
&\delta \frac{1}{m(B(0,R+R^{\kappa}))} \int_{B(0,R)} w^2\,\mathrm{d}x \\
&\quad \leq C \cdot R^{-1} + C_{\theta}\|w\|_{\infty} \cdot \frac{1}{m(B(0,R+R^{\kappa}))} \int_{B(0,R+R^{\kappa})} (|h_2(x)| + |\varphi(x,v)|)\,\mathrm{d}x,
\end{aligned} \tag{4.8}$$

where C is a positive constant.

Since h2∈PAP0(RN), we have

$$\lim_{R\to\infty} \frac{1}{m(B(0,R+R^{\kappa}))} \int_{B(0,R+R^{\kappa})} |h_2(x)|\,\mathrm{d}x = 0. \tag{4.9}$$

By Lemma 2.2(2), the range Rυ of υ is relatively compact. We know from Lemma 2.5 that g(x,z) is uniformly continuous in $\mathbb{R}^N \times \overline{\mathcal{R}_\upsilon}$. Since f(x,z) is continuous in $z \in \overline{\mathcal{R}_\upsilon}$ and uniformly in $x \in \mathbb{R}^N$ then so is φ=f−g. Now by Lemma 3.9, we have

$$\lim_{R\to\infty} \frac{1}{m(B(0,R+R^\kappa))} \int_{B(0,R+R^\kappa)} |\varphi(x,\upsilon)|\, dx$$

$$\leq \lim_{R\to\infty} \frac{1}{m(B(0,R+R^\kappa))} \int_{B(0,R+R^\kappa)} \max_{z\in\mathcal{R}_\upsilon} |\varphi(x,z)|\, dx = 0. \tag{4.10}$$

It follows from (4.8)-(4.10) that

$$\lim_{R\to\infty} \frac{1}{m(B(0,R+R^\kappa))} \int_{B(0,R)} w^2\, dx = 0. \tag{4.11}$$

Note $\frac{m(B(0,R))}{m(B(0,R+R^\kappa))} = \frac{R^N}{(R+R^\kappa)^N} = \frac{1}{(1+R^{\kappa-1})} \to 1, \text{ as } R \to \infty.$ Then (4.11) implies that

$$\lim_{R\to\infty} \frac{1}{m(B(0,R))} \int_{B(0,R)} w^2\, dx = 0,$$

i.e., $w^2 \in \mathcal{PAP}_0(\mathbb{R}^N)$. From Lemma 3.4, it follows that w∈PAP0(RN). □

Remark 4.7 Theorem 4.6 can also be comprehended in the following way. If the nonlinear term g and forcing term h1 of (4.4) are perturbed by ergodic perturbations which maintain the monotonicity of nonlinear term and the existence of upper and lower solutions, then its almost periodic weak solution is also perturbed by an ergodic perturbation.

4.3 Pseudo almost periodic forced pendulum equation

Consider the equation

$$-\Delta u + \sum_{j=1}^{N} c_j \partial_j u - f(x) \cdot \sin u = h(x), \quad x \in \mathbb{R}^N, \tag{4.12}$$

where cj's are real constants, *f* and *h* both are pseudo almost periodic. If there exists an ϵ>0 such that $f(x) \geq \|h\|_\infty + \epsilon, \forall x \in \mathbb{R}^N$, then (4.12) has a unique pseudo almost periodic weak solution such that $\frac{\pi}{2} < u < \frac{3\pi}{2}$.

A special case of (4.12) was treated in [[14], Example 3.3] and [[15], Example 7.1(b)], where f(x)≡a, *h* is almost periodic and a>||h||∞.

Proof Let *g* and h1 are the almost periodic components of *f* and *h*, respectively. We need to consider the following almost periodic equation together:

$$-\Delta v + \sum_{j=1}^{N} c_j \partial_j v - g(x) \cdot \sin v = h_1(x), \quad x \in \mathbb{R}^N. \tag{4.13}$$

It follows from Lemma 3.6 that *g* and h1 also satisfy the relationship that g(x)≥||h1||∞+ϵ, ∀x∈RN.

If δ>0 is sufficient small (<ϵ), then $\alpha = \frac{\pi}{2} + \delta$ and $\beta = \frac{3\pi}{2} - \delta$ are lower and upper solutions of both (4.12) and (4.13).

Since −∂u[f(x)·sinu]=−f(x)·cosu is strictly positive when $u \in [\frac{\pi}{2} + \delta, \frac{3\pi}{2} - \delta]$.

Now we can apply Theorem 4.6 and get the conclusion that (4.12) has a unique pseudo almost periodic weak solution *u* such that

$\frac{\pi}{2} < u < \frac{3\pi}{2}$ and its almost periodic component is a weak solution of (4.13). □

ACKNOWLEDGEMENTS

The authors acknowledge support from the NSF of China (no. 11071048).

COMPETING INTERESTS

The authors declare that they have no competing interests.

AUTHORS' CONTRIBUTIONS

All authors contributed equally in this article. They read and approved the final manuscript.

REFERENCES

1. Zhang CY: Integration of vector-valued pseudo almost periodic functions. *Proc. Am. Math. Soc.* 1994, 121: 167-174.

2. Zhang CY: Pseudo almost periodic solutions of some differential equations. *J. Math. Anal. Appl.* 1994, 151: 62-76.

3. Zhang CY: Pseudo almost periodic solutions of some differential equations, II. *J. Math. Anal. Appl.* 1995, 192: 543-561.

4. Li HX, Huang FL, Li JY: Composition of pseudo almost-periodic functions and semilinear differential equations. *J. Math. Anal. Appl.* 2001, 255: 436-446.

5. Cieutat P, Fatajou S, N'Guérékata GM: Composition of pseudo almost periodic and pseudo almost automorphic functions and applications to evolution equations. *Appl. Anal.* 2010, 89: 11-27.

6. Dads EA, Ezzinbi K: Existence of positive pseudo almost periodic solution for a class of functional equations arising in epidemic problems. *Cybern. Syst. Anal.* 1994, 30: 900-910.

7. Xu B, Yuan R: The existence of positive almost periodic type solutions for some neutral nonlinear integral equation. *Nonlinear Anal., Theory Methods Appl.* 2005, 60: 669-684.

8. Diagana T: *Pseudo Almost Periodic Functions in Banach Spaces.* Nova Science Publishers, New York; 2007.

9. Bochner S: Abstrakte fastperiodische funktionen. *Acta Math.* 1933, 61: 149-184.

10. Sibuya Y: Almost periodic solutions of Poisson's equation. *Proc. Am. Math. Soc.* 1971, 28: 195-198

11. Sell GR: Almost periodic solutions of linear partial differential equations. *J. Math. Anal. Appl.* 1973, 42: 302-312.

12. Zaidman S: *Almost Periodic Functions in Abstract Spaces.* Pitman, Boston; 1985.

13. Pankov A: *Bounded and Almost Periodic Solutions of Nonlinear Operator Differential Equations.* Kluwer Academic, Dordrecht; 1990. English Edition

14. Fournier G, Szulkin A, Willem M:Semilinear elliptic equations in Rn with almost periodic or unbounded forcing term. *SIAM J. Math. Anal.*1996, 27: 1653-1660.

15. N'Guérékata GM, Pankov A: Almost periodic elliptic equations: sub- and super-solutions. Operator Theory: Advances and Applications 228. In *Operator Theory, Pseudo-Differential Equations, and Mathematical Physics.* Birkhäuser, Basel; 2013:275-292.

16. Shubin MA: Differential and pseudodifferential operators in spaces of almost periodic functions. *Math. USSR Sb.* 1974, 24(4):547-573.

17. Shubin MA: Almost periodic functions and partial differential operators. *Russ. Math. Surv.* 1978, 33(2):1-52.

6

LAYER SOLUTIONS FOR A CLASS OF SEMILINEAR ELLIPTIC EQUATIONS INVOLVING FRACTIONAL LAPLACIANS

Yan Hu

College of Mathematics and Econometrics, Hunan University, Changsha, 410082, China

ABSTRACT

This paper is concerned with the nonlinear equation involving the fractional Laplacian: $(-\Delta)^s v(x) = b(x)f(v(x)), x \in \mathbb{R}$, where $s \in (0,1)$, $b:\mathbb{R} \to \mathbb{R}$ is a periodic, positive, even function and $-f$ is the derivative of a double-well potential G. That is $G \in C^{2,\gamma}$ $(0 < \gamma < 1)$, $G(1) = G(-1) < G(\tau)$ $\forall\tau$. $\in (-1,1)$, $G'(-1) = G'(1) = 0$. We show the existence of layer solutions of the equation for s≥12 and for some odd nonlinearities by variational methods, which is a bounded solution having the limits ±1 at ±∞. Asymptotic estimates for layer solutions as |x|→+∞ and the asymptotic behavior of them as s↑1 are also obtained.

KEYWORDS

fractional Laplacian, layer solutions, existence, local minimizers

1 INTRODUCTION

In this paper we study the fractional Laplacian

$$(-\triangle)^s v(x) = b(x) f\left(v(x)\right), \quad x \in \mathbb{R}, \tag{1.1}$$

where $s \in (0,1)$, and $(-\triangle)^s$ is the fractional Laplacian defined by

$$(-\triangle)^s v = C_s \text{ P.V.} \int_{\mathbb{R}} \frac{v(x) - v(y)}{|x-y|^{1+2s}} \, dy.$$

Here P.V. stands for the Cauchy principle value and Cs is a positive constant multiplier depending only on *s*.

The fractional Laplacian is a nonlocal operator which can be localized as

$$\begin{cases} -\operatorname{div}(y^a \nabla u) = 0 & \text{in } \mathbb{R}^2_+, \\ -\lim_{y \downarrow 0^+} y^a \frac{\partial u}{\partial y} = \frac{1}{d_s} b(x) f(u) & \text{on } \partial\mathbb{R}^2_+, \end{cases} \tag{1.2}$$

where $a = 1 - 2s \in (-1,1)$, $d_s = 2^{2s-1} \frac{\Gamma(s)}{\Gamma(1-s)}$ and $u(x,0) = v(x)$.
Moreover $u(\cdot,\cdot)$ can be expressed by a Poisson kernel,

$$u(x,y) = P_s(\cdot,y) * v = p_s \int_{\mathbb{R}} \frac{y^{2s}}{\left(|z|^2 + y^2\right)^{\frac{1+2s}{2}}} v(x-z) \, dz \quad \text{for every } y > 0,$$

which is called the *s*-extension of *v*. ps is a positive constant depending only on *s*. For more details as regards the fractional Laplacian, readers can refer to [1–7] and the references therein.

In view of the celebrated De Giorgi conjecture (see [8–10]), Cabré and Sire [2, 3] considered layer solutions of the nonlocal equation

$$(-\Delta)^s v = f(v) \quad \text{in } \mathbb{R}. \tag{1.3}$$

The necessary and sufficient conditions for the existence of one-dimensional layer solutions were given as

$$G(1) = G(-1) < G(s) \quad \forall s \in (-1,1), \qquad G'(1) = G'(-1) = 0,$$

where G′=−f. All these were obtained by a Hamiltonian equality and a Modica-type estimate for layer solutions. By the sliding method, the layer solution of (1.3) was proved to be the unique local minimizer which increases in *x* with values varying from −1 to 1. The regularity, Hopf principle, maximum principle as well as a Harnack inequality for (1.3) or for its extension equation (1.2) (in this case b=1) were given. Some of them will be used in our paper.

If *b* is not a constant and is periodic, the perturbed equation (1.1) becomes complicated. The aim of this paper is to study the layer solution of (1.1) with periodic perturbed nonlinearity.

Definition 1.1 A function $v \in (L^\infty \cap C^\beta)(\mathbb{R})$ $(0 < \beta < 1)$ is said to be a layer solution of (1.1), if *v* solves (1.1),

$$(-\Delta)^s v(x) = b(x) f(v(x)), \quad x \in \mathbb{R}$$

and

$$\lim_{x \to \pm\infty} v(x) = \pm 1.$$

Definition 1.2 A function $u \in L^\infty(\mathbb{R}^2_+) \cap C^\beta(\overline{\mathbb{R}^2_+})$ is said to be a layer solution of (1.2), if *u* solves (1.2),

$$\begin{cases} -\operatorname{div}(y^a \nabla u) = 0 & \text{in } \mathbb{R}^2_+, \\ -\lim_{y\downarrow 0^+} y^a \frac{\partial u}{\partial y} = \frac{1}{d_s} b(x) f(u) & \text{on } \partial\mathbb{R}^2_+ \end{cases}$$

and

$$\lim_{x\to\pm\infty} u(x,0) = \pm 1.$$

Namely, u(x,0) is the corresponding layer solution of (1.1).

Different from the unperturbed case (1.3), the inhomogeneous term b(x)f(u) depends explicitly on x in (1.1) and (1.2); the sliding method cannot be used and layer solutions of them have no monotonicity in the direction of x. The method for obtaining layer solutions in [2] and [3] cannot be used in our case directly; some difficulties need to be solved.

In the paper, we consider the extension problem (1.2). Obviously, (1.2) has a variational structure.

Denote

$\Omega \subset \mathbb{R}^2_+$, a bounded Lipschitz domain,

$B_R(x,y) \subset \mathbb{R}^2$, a ball centered at $(x,y) \in \mathbb{R}^2$ with radius R,

$B^+_\epsilon(x,0) = B_\epsilon(x,0) \cap \mathbb{R}^2_+$,

$\partial^0\Omega = \left\{(x,0) \in \partial\Omega \cap \partial\mathbb{R}^2_+ \,|\, \exists\epsilon > 0, B^+_\epsilon(x,0) \subset \Omega\right\}$,

$\partial^+\Omega = \overline{\partial\Omega \cap \mathbb{R}^2_+}$.

For $u \in H^1(y^a,\Omega)$, the norm is

$$\|u\|_{H^1(y^a,\Omega)} = \left(\int_\Omega y^a |\nabla u|^2 \, dx\, dy\right)^{\frac{1}{2}} + \left(\int_\Omega y^a |u|^2 \, dx\, dy\right)^{\frac{1}{2}}.$$

The energy functional of u on Ω is given by

$$\mathcal{E}(u,\Omega) = d_s \int_{\Omega} \frac{y^a}{2} |\nabla u|^2 \, dx\, dy + \int_{\partial^0 \Omega} b(x) G(u(x,0)) \, dx. \tag{1.4}$$

We state our main results in the following.

We show, via a Liouville result, the existence of layer solutions of (1.1) for s≥12 and for some odd nonlinearities.

Theorem 1.1 *Let* $s \geq \frac{1}{2}$*Assume that* b, $f \in C^{1,\gamma}(\mathbb{R})$ $(0 < \gamma < 1)$: :

*(1)*b:R→R *is* 1*-periodic, even, not constant and positive; denote* b¯= maxRb *and* b−=minRb;

*(2)*f(−τ)=−f(τ) *for any* τ∈[−1,1], f(−1)=f(1)=f(0)=0, f>0 *in* (0,1) *and* f<0 *in* (−1,0).

Obviously, if G′=−f,

$$G(-1) = G(1) < G(\tau) \quad \text{for } \tau \in (-1,1), \qquad G'(1) = G'(-1) = 0.$$

There exists a layer solution v∈C2,β(R) *(for some* 0<β<1) *of* (1.1):

$$\begin{cases} (-\partial_{xx})^s v(x) = b(x) f(v(x)) & \text{in } \mathbb{R}, \\ v \to \pm 1 & \text{as } x \to \pm\infty. \end{cases} \tag{1.5}$$

In addition, v is odd.

Furthermore we obtain asymptotic estimates of the layer solutions of (1.1) by comparing with a layer solution of the unperturbed equation (1.3).

Theorem 1.2 *Let* b∈C1,γ∩L∞ *is positive. Let* $b \in C^{1,\gamma} \cap L^\infty$ $f \in C^{1,\gamma}(\mathbb{R})$ $(\gamma > \max(0, 1-2s))$ *satisfy*

*(i)*G(−1)=G(1)<G(τ) *for* τ∈(−1,1), G′(1)=G′(−1)=0;

*(ii)*G″(1)>0, G″(−1)>0.

If v *is a layer solution of* (1.1), *then the following asymptotic estimates hold*:

$$cx^{-2s} \leq |1 - v| \leq Cx^{-2s} \quad for\ x > 1, \tag{1.6}$$

$$c|x|^{-2s} \leq |1 + v| \leq C|x|^{-2s} \quad for\ x < -1 \tag{1.7}$$

for some constants 0<c<C.

Finally we investigate the asymptotic behavior of vs as s↑1 and obtain a local elliptic equation, which is stated as follows.

Theorem 1.3 *Let* s∈[12,1). *Let* {vsk} *be a sequence of layer solutions of* (1.1) *in Theorem* 1.1. *Then, there exists a subsequence denoted again by* {vsk} *converging locally uniformly to a function* v1∈C2(R) *as* sk↑1, *which is also a layer solution of the local elliptic equation*

$$\begin{cases} -v^1_{xx}(x) = b(x)f(v^1(x)) & in\ \mathbb{R}, \\ \lim_{x\to\pm\infty} v^1(x) = \pm 1. \end{cases} \tag{1.8}$$

In addition,

$$\frac{1}{2}\left(v^1_x\right)^2 = b(x)\left\{G\left(v^1(x)\right) - G(1)\right\} + \int_x^{+\infty} b'(t)\left\{G\left(v^1(t)\right) - G(1)\right\} dt. \tag{1.9}$$

For convenience of the presentation we will use *C* for a general positive constant; such a *C* is usually different in different contexts.

2 SOME PRELIMINARIES AND PROPERTIES

In this paper, we mainly study the extension equation (1.2). To make our problems clear, we present several properties of layer solutions.

Lemma 2.1 *Let u be a bounded solution of* (1.2),

$$\begin{cases} -\operatorname{div}(y^a \nabla u) = 0 & in\ \mathbb{R}^2_+, \\ -\lim_{y\downarrow 0^+} y^a \frac{\partial u}{\partial y} = \frac{1}{d_s} b(x) f(u) & on\ \partial\mathbb{R}^2_+ \end{cases}$$

and

$$\lim_{x\to\pm\infty} u(x,0) = L^{\pm} \tag{2.1}$$

with two constants L±. *Then,*

(1) $f(L^+) = f(L^-) = 0;$ (2.2)

(2) $\lim_{x\to\pm\infty} u(x,y) = L^{\pm}$ (2.3)

for every y≥0;

(3) $\left\| u - L^{\pm} \right\|_{L^\infty(B_R^+(x,0))} \to 0 \quad as\ x \to \pm\infty;$ (2.4)

(4) $\| \nabla_x u \|_{L^\infty(B_R^+(x,0))} + \left\| y^a \frac{\partial u}{\partial y} \right\|_{L^\infty(B_R^+(x,0))} \to 0 \quad as\ x \to \pm\infty.$ (2.5)

Proof Our proof uses the invariance of the problem under periodic translations in *x* and a compactness argument.

Denote un(x,y)=u(x+n,y) for n∈Z. Since *b* is 1-periodic, un still satisfies the equations

$$\begin{cases} -\operatorname{div}(y^a \nabla u^n) = 0 & \text{in } \mathbb{R}^2_+, \\ -\lim_{y\downarrow 0^+} y^a \frac{\partial u^n}{\partial y} = \frac{1}{d_s} b(x) f(u^n(x,0)) & \text{on } \partial\mathbb{R}^2_+. \end{cases} \tag{2.6}$$

By regularity results in [2] and [5], we see that up to a subsequence,

$$u^n \to u^{\pm\infty} \quad \text{in } C^0_{\text{loc}}\left(\overline{\mathbb{R}^2_+}\right),$$
$$\nabla_x u^n \to \nabla_x u^{\pm\infty} \quad \text{in } C^0_{\text{loc}}\left(\overline{\mathbb{R}^2_+}\right),$$
$$y^a \frac{\partial u^n}{\partial y} \to y^a \frac{\partial u^\infty}{\partial y} \quad \text{in } C^0_{\text{loc}}\left(\overline{\mathbb{R}^2_+}\right)$$

as $n \to \pm\infty$. Then $u^{\pm\infty}$ solves the equations

$$\begin{cases} -\operatorname{div}(y^a \nabla u^{\pm\infty}) = 0 & \text{in } \mathbb{R}^2_+, \\ -\lim_{y\downarrow 0^+} y^a \frac{\partial u^{\pm\infty}}{\partial y} = \frac{1}{d_s} b(x) f(u^{\pm\infty}) & \text{on } \partial\mathbb{R}^2_+, \end{cases} \tag{2.7}$$

and it follows that $u^{\pm\infty}(x,0) \equiv L^\pm$ for every $x \in \mathbb{R}$.

Consider the Dirichlet problem

$$\begin{cases} -\operatorname{div}(y^a \nabla u^{\pm\infty}) = 0 & \text{in } \mathbb{R}^2_+, \\ u^{\pm\infty}(x,0) \equiv L^\pm & \text{on } \partial\mathbb{R}^2_+. \end{cases} \tag{2.8}$$

$u^{\pm\infty} \equiv L^\pm$ is the unique solution of (2.8) by Corollary 3.5 in [2]. As a consequence, (2.2) and (2.5) are obvious. □

The following lemma is a necessary condition for a local minimizer of the energy functional $\mathcal{E}$.

Lemma 2.2 *Let u be a local minimizer of the energy functional $\mathcal{E}$ under perturbations in* $[-1,1]$. *That is, for any bounded Lipschitz domain* $\Omega \subset \mathbb{R}^2_+$ *and for any* $\xi \in H^1(y^a,\Omega)$ *having compact support in* $\Omega \cup \partial^0\Omega$ *such that* $u+\xi \in [-1,1]$,

$$\mathcal{E}(u,\Omega) \le \mathcal{E}(u+\xi,\Omega).$$

Let

$$\lim_{x\to\pm\infty} u(x,0) = \pm 1. \tag{2.9}$$

Then

$$G(1) = G(-1) \le G(\tau) \quad \textit{for all } \tau \in (-1,1). \tag{2.10}$$

Proof To show (2.10), it is sufficient to prove that $G(1)\le G(\tau)$ and $G(-1)\le G(\tau)$ for all $\tau\in[-1,1]$. Suppose $G(\tau_0)< G(1)$ for some point $\tau_0\in[-1,1]$ by contradiction. For simplicity, assume that $G(\tau_0)=0$ by adding a constant.

By (2.9),

$$\liminf_{l\to+\infty} \mathcal{E}\big(u, B_R^+(l,0)\big) \ge \liminf_{l\to+\infty} \int_{\partial^0 B_R^+(l,0)} b(x) G\big(u(x,0)\big) \ge 2\underline{b}\varepsilon R \tag{2.11}$$

for some $\varepsilon>0$.

Let ξ_R be a cut-off function with values in $[0,1]$,

$$\xi_R = \begin{cases} 1 & \text{in } B_{(1-\eta)R}, \\ 0 & \text{in } \mathbb{R}^{n+1}\setminus B_R, \end{cases}$$

where $\eta\in(0,1)$ will be specified later, and $|\nabla \xi_R| \le \frac{1}{\eta R}$.

Define $\xi_{R,l}(x,y) = \xi_R(x-l,y)$. Let $w = \tau_0\xi_{R,l} + (1-\xi_{R,l})u$, then $w = u$ on $\partial^+ B_R^+(l,0)$ and $w \equiv \tau_0$ in $B_{(1-\eta)R}^+(l,0)$. We have

$$\limsup_{l\to+\infty} \mathcal{E}_{B_R^+(l,0)}(w) = \limsup_{l\to+\infty} \left\{ d_s \int_{B_R^+(l,0)} \frac{y^a}{2} \left|(1-\xi_{R,l})\nabla u + (\tau_0 - u)\nabla \xi_{R,l}\right|^2 + \int_{\partial^0 B_R^+(l,0)} b(x_1)G(w) \right\}$$
$$\leq 2d_s \int_{B_R^+} y^a |\nabla \xi_{R,l}|^2 + 2\bar{b} \max_{[-1,1]} G \cdot \eta R$$
$$\leq \frac{Cd_s R^a}{\eta^2} + 2\bar{b} \max_{[-1,1]} G \cdot \eta R. \tag{2.12}$$

We use (2.5) in the first inequality above.

Having chosen $\eta = \frac{\underline{b}\varepsilon}{2\bar{b}\max_{[-1,1]} G}$, by (2.11) and (2.12),

$$\limsup_{l\to+\infty} \mathcal{E}\left(w, B_R^+(l,0)\right) < \liminf_{l\to+\infty} \mathcal{E}\left(u, B_R^+(l,0)\right)$$

for large R>1. This contradiction leads to G(1)≤G(τ) for all τ∈[−1,1]. By the same discussion, G(−1)≤G(τ) for all τ∈[−1,1]. Thus we complete the proof. □

As in [2], we construct a Hamiltonian equality which will be used in the proof of Theorem 1.3. For this purpose a lemma is in order, for whose proof see Lemma 5.1 in [2].

Lemma 2.3 *Let* $u \in L^\infty(\mathbb{R}^2_+)$ *be a solution of* (1.2). *Then for every* x $\in \mathbb{R}$, $\int_0^\infty y^a |\nabla u|^2 \, dy < \infty$. *In addition, the integral can be differentiated with respect to* x∈R *under the integral sign. We have*

$$\lim_{M\to+\infty} \int_M^\infty y^a |\nabla u|^2 \, dy = 0 \tag{2.13}$$

uniformly in x∈R. *If u is a layer solution of* (1.2),

$$\lim_{|x|\to+\infty} \int_0^\infty y^a |\nabla u|^2 \, dy = 0. \tag{2.14}$$

Proposition 2.1 (Hamiltonian equality)

Let u be a layer solution of (1.2) *for* a∈(−1,12), *i.e.,*

$$\begin{cases} -\operatorname{div}(y^a \nabla u) = 0 & in\ \mathbb{R}^2_+, \\ -\lim_{y\downarrow 0^+} y^a \frac{\partial u}{\partial y} = \frac{1}{d_s} b(x) f(u) & on\ \partial\mathbb{R}^2_+, \\ u(x,0) \to \pm 1 & as\ x \to \infty. \end{cases}$$

For every x∈R, *the Hamiltonian equality holds*:

$$d_s \int_0^\infty \frac{y^a}{2} \{u_x^2 - u_y^2\}\, dy = b(x)\{G(u(x,0)) - G(1)\} + \int_x^\infty b'(t)\{G(u(t,0)) - G(1)\}\, dt. \tag{2.15}$$

As a consequence,

$$\int_{-\infty}^{+\infty} b'(x)\{G(u(x,0)) - G(1)\}\, dx = 0. \tag{2.16}$$

Proof We note that the integral in (2.16) is well defined since $s = \frac{1-a}{2} \in (\frac{1}{4}, 1)$ and

$$G(u(x,0)) - G(1) = \frac{G''(t)}{2}(u(x,0)-1)^2 = O(|x|^{-4s}) \quad \text{as } |x| \to \infty,$$

where *t* is some point between u(x,0) and 1.

By Lemma 2.3, the left integral in (2.15) can be differentiated with respect to *x*,

$$\frac{d}{dx}d_s\int_0^{\infty}\frac{y^a}{2}\{u_x^2-u_y^2\}\,dy = d_s\int_0^{\infty}y^a\{u_xu_{xx}-u_yu_{yx}\}\,dy$$
$$= d_s\int_0^{\infty}y^a\left\{u_{xx}+u_{yy}+\frac{a}{y}u_y\right\}u_x\,dy + d_s\lim_{y\to 0^+}y^a u_y u_x$$
$$= -b(x)f\big(u(x,0)\big)u_x(x,0).$$

In the second equality above we use the fact that $\lim_{y\to\infty} y^a u_y u_x = 0$ (see [2]). We have

$$\frac{d}{dx}\left\{b(x)\big(G(u(x,0))-G(1)\big)+\int_x^{+\infty}b'(t)\big(G(u(t,0))-G(1)\big)\,dx\right\}$$
$$= -b(x)f\big(u(x,0)\big)u_x(x,0).$$

Thus,

$$d_s\int_0^{\infty}\frac{y^a}{2}\{u_x^2-u_y^2\}\,dy \equiv b(x)\{G(u(x,0))-G(1)\}$$
$$+\int_x^{+\infty}b'(t)\{G(u(t,0))-G(1)\}\,dt + C. \tag{2.17}$$

Let $x\to+\infty$, the left of (2.17) converging to zero by (2.14); thus $C=0$ and (2.15) is proved. Letting $x\to-\infty$, (2.16) is also obtained. □

To study asymptotic estimates of layer solutions of (1.1), we recall an explicit layer solution of the unperturbed problem (1.3).

Lemma 2.4 ([3], Theorem 3.1) *Let* $s\in(0,1)$. *For every* $t>0$, *the* $C\infty$ *function*

$$v_s^t(x) = \operatorname{sign}(x)\frac{2}{\pi}\int_0^{\infty}\frac{\sin(z)}{z}e^{-t(\frac{z}{|x|})^{2s}}\,dz \tag{2.18}$$

is the layer solution to the fractional equation

$$(-\partial_{xx})^s v_s^t = f_s^t(v_s^t) \quad in\ \mathbb{R}, \tag{2.19}$$

for a nonlinearity $f_s^t \in C^1([-1,1])$ *which is odd and twice differentiable in* [−1,1] *and which satisfies*

$$f_s^t(0) = f_s^t(1) = 0, \quad f_s^t > 0 \; in \; (0,1), \qquad (f_s^t)'(\pm 1) = -\frac{1}{t}.$$

In addition, the following limits exist:

$$\lim_{|x|\to\infty} |x|^{1+2s}\left(\partial_x v_s^t\right)(x) = t\frac{4s}{\pi}\sin(\pi s)\Gamma(2s) > 0 \tag{2.20}$$

and, as a consequence,

$$\lim_{x\to\pm\infty} |x|^{2s}\left|\left(v_s^t\right)(x) \mp 1\right| = t\frac{2}{\pi}\sin(\pi s)\Gamma(2s) > 0. \tag{2.21}$$

3 EXISTENCE AND ASYMPTOTIC ESTIMATES

To prove the existence of layer solutions, we introduce a Liouville result where a≤0 is required. This is the reason why we restrict ourselves to the case s≥12 in Theorem 1.1.

Proposition 3.1 *Let* a≤0. *Suppose u is a bounded nonnegative function which satisfies weakly the problem*

$$\begin{cases} -\operatorname{div}(y^a \nabla u) \le 0 & in \; \mathbb{R}^2_+, \\ -\lim_{y\downarrow 0^+} y^a \frac{\partial u}{\partial y} \le 0 & on \; \partial\mathbb{R}^2_+. \end{cases} \tag{3.1}$$

Then $u \equiv C \; a.e. \; in \; \mathbb{R}^2_+.$

Proof Since a≤0, Ra≤1 for R>1. Let ξ be a smooth function with values in [0,1], ξ=1 in BR and ξ=0 outsidc of B2R, |∇ξ|≤CR 1. Multiplying (3.1) with uξ2 and integrating by parts, we have that

$$\int_{B_R^+} y^a|\nabla u|^2 \leq \int_{B_{2R}^+} y^a|\nabla u|^2\xi^2 \leq 2\int_{\mathbb{R}_+^2} y^a\xi u|\nabla u||\nabla\xi|$$

$$\leq 2\left\{\int_{B_{2R}^+\setminus B_R^+} y^a|\nabla u|^2\xi^2\right\}^{\frac{1}{2}}\left\{\int_{B_{2R}^+\setminus B_R^+} y^a|\nabla\xi|^2u^2\right\}^{\frac{1}{2}}$$

$$\leq C\left\{\int_{B_{2R}^+\setminus B_R^+} y^a|\nabla u|^2\xi^2\right\}^{\frac{1}{2}}(RR^{1+a}R^{-2})^{\frac{1}{2}}$$

$$\leq C\left\{\int_{B_{2R}^+\setminus B_R^+} y^a|\nabla u|^2\xi^2\right\}^{\frac{1}{2}}.$$

Thus $\int_{\mathbb{R}_+^2} y^a|\nabla u|^2 \leq C$ for some constant C independent of R. Let $\int_{B_{2R}^+\setminus B_R^+} y^a|\nabla u|^2\xi^2 \to 0.$. We deduce that $\int_{\mathbb{R}_+^2} y^a|\nabla u|^2 = 0$ and $u \equiv C$ a.e. in $\mathbb{R}_+^2$. .

Next we prove an existence result about the local minimizer of $\mathcal{E}$.

Lemma 3.1 *Let Ω⊂R2+ be a bounded Lipschitz domain. Let* $w_0 \in C^0(\overline{\Omega}) \cap H^1(y^a, \Omega)$ *be a given function with |w0|≤1; bis a bounded positive function.*

Suppose that

$$f(1) \leq 0 \leq f(-1),$$

the energy functional $\mathcal{E}(u,\Omega)$*admits a minimizer* $u \in C_{w_0,a} = \{w \in H^1(y^a,\Omega), -1 \leq w \leq$ 1 *a.e. in* $\Omega, w = w_0$ *on* $\partial^+\Omega$ in the weak sense}, *which solves weakly*

$$\begin{cases} -\operatorname{div}(y^a\nabla u) = 0 & \textit{in } \Omega, \\ -\lim_{y\downarrow 0^+} y^a\frac{\partial u}{\partial y} = \frac{1}{d_s}b(x)f(u(x,0)) & \textit{on } \partial^0\Omega, \\ u = w_0 & \textit{on } \partial^+\Omega. \end{cases} \tag{3.2}$$

Moreover, u is a stable solution of (3.2), *i.e.,*

$$d_s\int_\Omega y^a|\nabla\xi|^2\,dx\,dy - \int_{\partial^0\Omega} b(x)f'(u)\xi^2\,dx \geq 0, \tag{3.3}$$

for every ξ∈H1(Ω,ya) *such that* ξ≡0 *on* ∂+Ω $\xi \in H^1(\Omega, y^a)$ *such that* $\xi \equiv 0$ *on* $\partial^+\Omega$ *in the weak sense.*

Proof Consider the set $H_{w_0,a}(\Omega) = \{w \in H^1(y^a, \Omega), w \equiv w_0 \text{ on } \partial^+\Omega$ in the weak sense}⊃ $C_{w_0,a}$, $H_{w_0,a}(\Omega) \neq \emptyset$ since $w_0 \in H_{w_0,a}(\Omega)$. Denote

$$\widetilde{f} = \begin{cases} f(1) & \text{if } t \geq 1, \\ f & \text{if } -1 < t < 1, \\ f(-1) & \text{if } t \leq -1, \end{cases}$$

and $\widetilde{G} = -\int_0^u \widetilde{f}$. Up to an additive constant, G˜=G in [−1,1].

Consider the energy functional

$$\widetilde{\mathcal{E}}(u, \Omega) = d_s \int_\Omega \frac{y^a}{2} |\nabla u|^2 \, dx \, dy + \int_{\partial^0 \Omega} b(x) \widetilde{G}(u(x,0)) \, dx. \tag{3.4}$$

If E˜ has an absolute minimizer *u* in Cw0,a(Ω), the statement of Lemma 3.1 is proved.

For every function $w \in H_{w_0,a}(\Omega)$, $w - w_0 \in H^1(y^a, \Omega)$ and vanishes on ∂+Ω in the weak sense. We can extend w−w0 in R2+ by zeroes outside of Ω and $w - w_0 \in H^1(y^a, \mathbb{R}^2_+)$. By the trace theorem and the Sobolev imbedding theorem (see [7, 11,12]),

$$H^1(y^a, \mathbb{R}^2_+) \hookrightarrow L^p(\mathbb{R})$$

for $p = \frac{2}{1-2s}$ if $s < \frac{1}{2}$ or for any $1 \leq p < \infty$ if $s \geq \frac{1}{2}$.

Moreover, $H^1(y^a, \mathbb{R}^2_+) \hookrightarrow\hookrightarrow L^2(\partial^0\Omega)$.

Since G˜ has linear growth at infinity, E˜ is well defined, bounded below and coercive in Hw0,a. There exists an absolute minimizer u∈Hw0,a. By the first order variation, we have

$$\begin{cases} -\operatorname{div}(y^a \nabla u) = 0 & \text{in } \Omega, \\ -\lim_{y\downarrow 0^+} y^a \frac{\partial u}{\partial y} = \frac{1}{d_s} b(x_1)\widetilde{f}(u(x,0)) & \text{on } \partial^0\Omega. \end{cases} \tag{3.5}$$

Multiply (u−1)+ with (3.5) and integrate in Ω,

$$d_s \int_\Omega y^a |\nabla(u-1)^+|^2\, dx\, dy - \int_{\partial^0\Omega} b(x) f(1)(u-1)^+\, dx = 0.$$

Since $f(1) \le 0$, $\int_\Omega y^a|\nabla(u-1)^+|^2\, dx\, dy \le 0$. Thus $(u-1)^+ \equiv 0$ a.e. in Ω, *i.e.*, u≤1 a.e. in Ω. Similarly we also get u≥−1 a.e. in Ω. Hence $u \in C_{w_0,a}(\Omega)$.. (3.2) follows from (3.5), and (3.3) comes from the second order variation of ℰ. □

Remark 3.1 Suppose that *b* is an even function, *f* and w0 are odd with respect to *x*, with a slight modification we can also show that there is an odd minimizer in the admissible set $\{w \in C_{w_0,a} | w(-x,y) = -w(x,y) \text{ for every } y \ge 0\}$.

Now we start to show the existence of layer solutions of (1.2).

Theorem 3.1 *Let* $s \ge \frac{1}{2}$. *Let* $b \in (C^{1,\gamma} \cap L^\infty)(\mathbb{R})$ *and* $f \in C^{1,\gamma}(\mathbb{R})$ $(0 < \gamma < 1)$:

*(a)*b:R→R *is an even positive function,* $b(x+1) = b(x)\ \forall x \in \mathbb{R}$,,

(b) $f(-\tau) = -f(\tau)$ *for any* $\tau \in [-1,1]$, $f(-1) = f(1) = f(0) = 0$, $f > 0$ *in* $(0,1)$ *and* $f < 0$ *in* $(-1,0)$.

Then there exists a layer solution u of (1.2) *in* $\mathbb{R}^2_+$:

$$\begin{cases} -\operatorname{div}(y^a \nabla u) = 0 & \text{in } \mathbb{R}^2_+, \\ -\lim_{y\downarrow 0^+} y^a \frac{\partial u}{\partial y} = \frac{1}{d_s} b(x) f(u(x,0)) & \text{on } \partial\mathbb{R}^2_+, \end{cases} \tag{3.6}$$

which is odd with respect to x, i.e., u(−x,y)=−u(x,y), *and, for every* y≥0,

$$\lim_{x \to \pm\infty} u(x,y) = \pm 1. \tag{3.7}$$

Furthermore, u is a local minimizer of the energy functional $\mathcal{E}$ under odd perturbations in [−1,1], *and it is stable in the sense that*

$$d_s \int_{\mathbb{R}^2_+} y^a |\nabla \xi|^2 \, dx \, dy - \int_{\mathbb{R}} b(x) f'(u(x,0)) \xi^2 \, dx \geq 0 \tag{3.8}$$

for every function $\xi \in C^1(\overline{\mathbb{R}^2_+})$ *with compact support in* $\overline{\mathbb{R}^2_+}$, $\xi(-x,y) = -\xi(x,y)$ *and* u+ξ∈[−1,1].

Proof The proof is divided into three parts. For simplicity, we make G(1)=G(−1)=0 by adding a constant.

Step 1. We show that there exists a solution with values in [−1,1] of (3.6) which is odd with respect to the variable *x* for every y≥0.

Let $Q_R = [-R,R] \times [0,R]$ and $w_0 = \frac{\arctan x}{\arctan R}$. Define the admissible set

$$C_{w_0,a,o} = \{ w \in C_{w_0,a}(Q_R), \forall y \geq 0, w(-x,y) = -w(x,y) \}.$$

By Remark 3.1, there is a minimizer uR in Cw0,a,o,

$$\begin{cases} -\operatorname{div}(y^a \nabla u_R) = 0 & \text{in } Q_R, \\ -\lim_{y \downarrow 0^+} y^a \frac{\partial u_R}{\partial y} = \frac{1}{d_s} b(x) f(u_R(x,0)) & \text{on } \partial^0 Q_R, \\ u_R = w_0 & \text{on } \partial^+ Q_R. \end{cases} \tag{3.9}$$

Define

$$u_R := \begin{cases} u_R(-x,y) & \text{if } u_R(x,y) < 0 \text{ and } x > 0, \\ u_R(x,y) & \text{if } u_R(x,y) \geq 0 \text{ and } x > 0 \end{cases}$$

and uR(x,y):=−uR(−x,y) for x≤0. Thus uR≥0 for x>0 and y≥0. Obviously uR is still a minimizer of E(·,QR).

By the regularity results in [2], $u_R, \nabla_x u_R, y^a \frac{\partial u_R}{\partial y} \in C^\beta(Q_R)$ for some $0 < \beta < 1$ and the continuous module is uniform bounded. Up to a subsequence, $u_R \to u$, $(u_R)_x \to u_x$ and $y^a \frac{\partial u_R}{\partial y} \to y^a \frac{\partial u}{\partial y}$ in $C^0(\overline{B_s^+})$ as $R \to \infty$. as $R \to \infty$ for all $R > s + 2$. By the canonical diagonal procedure, u solves

$$\begin{cases} -\operatorname{div}(y^a \nabla u) = 0 & \text{in } \mathbb{R}^2_+, \\ -\lim_{y \downarrow 0^+} y^a \frac{\partial u}{\partial y} = \frac{1}{d_s} b(x) f(u(x,0)) & \text{on } \partial\mathbb{R}^2_+, \\ u(-x,y) = -u(x,y) & \text{in } \overline{\mathbb{R}^2_+}, \end{cases} \tag{3.10}$$

and by the Hopf maximum principle $-1<u<1$.

Step 2. We show that there exists at least a subsequence $x_n \to \infty$ such that $u(x_n,0) \to 1$.

First we claim that u is a local minimizer under odd perturbations in $[-1,1]$. That is,

$$\mathcal{E}(u,\Omega) \le \mathcal{E}(w,\Omega)$$

for any $\Omega \subset \mathbb{R}^2_+$ and for any odd function $w \in H^1(y^a,\Omega)$ with $|w| \le 1$ and $w=u$ on $\partial^+\Omega$ in the weak sense.

Let $\xi \in C_c^1(B_s^+ \cup \partial^0 B_s^+)$ is odd with respect to x for every $y \ge 0$ and $u_R+\xi \in [-1,1]$. Since $-1<u_R<1$, $u_R+(1-\epsilon)\xi \in (-1,1)$ for $\epsilon \in (0,1)$. We have

$$\mathcal{E}\left(u_R, B_s^+\right) \le \mathcal{E}\left(u_R + (1-\epsilon)\xi, B_s^+\right) \quad \text{for } R > s+2.$$

Let $R \to \infty$, and

$$\mathcal{E}\left(u, B_s^+\right) \le \mathcal{E}\left(u + (1-\epsilon)\xi, B_s^+\right)$$

for every $s>0$ and $u+(1-\epsilon)\xi \in [-1,1]$. Our claim is proved.

If $w(-x,y)=-w(x,y)$,

$$\mathcal{E}\left(w,B_s^+\right)=2\mathcal{E}\left(w,B_s^{++}\right)=2\left\{d_s\int_{B_s^{++}}\frac{y^a}{2}|\nabla w|^2\,dx\,dy+\int_{\partial^0 B_s^{++}}b(x)G(w)\,dx\right\},$$

where $B_s^{++}=\{(x,y)\in B_s^+,x>0,y\geq 0\}.$. Therefore *u* is also a local minimizer of $\mathcal{E}$ in $\mathbb{R}_{++}^{n+1}=\{(x,y)\in\mathbb{R}_+^2,x>0,y\geq 0\}$ with perturbations in [−1,1], *i.e.*,

$$\mathcal{E}(u,\Omega)\leq\mathcal{E}(w,\Omega)$$

for any Ω⊂R2++ and for any w∈H1(ya,Ω) with |w|≤1 and w=u on ∂+Ω in weak sense.

Suppose u(xn,0)↛1 for any sequence xn→∞ by contradiction. |u(x,0)|<1−ϵ for some 0<ϵ<1 and x∈R. Hence0≤u(x,y)<1−ϵ for all x>0 and y≥0 by the fact that u(⋅,y)=Ps(⋅,y)∗u(⋅,0).

Let R>1. Let φR be a cut-off function with values 1 in $B_{(1-\eta)R}^+$and zeroes outside of B_R^+, $|\nabla\varphi_R|\leq\frac{C}{\eta R}$ for some 0<η<1determined later.

Denote φR=φR(|(x−l,y)|). Let $\varphi_R=\varphi_R(|(x-l,y)|)$. Let $w=1\cdot\varphi_R+(1-\varphi_R)u\in H^1(y^a,B_R^+(l,0))$, $w\equiv u$ on $\partial^+B_R(l,0)$. For l>R,

$$\begin{aligned}\mathcal{E}\left(w,B_R^+(l,0)\right)&=d_s\int_{B_R^+(l,0)}\frac{y^a}{2}\left|(1-\varphi_R)\nabla u+(1-u)\nabla\varphi_R\right|^2dx\,dy+\int_{\partial^0 B_R^+(l,0)}b(x)G(w)\,dx\\
&\leq d_s\int_{B_R^+(l,0)}\frac{y^a}{2}|\nabla u|^2\,dx\,dy+d_s\int_{B_R^+(l,0)}\frac{y^a}{2}|\nabla\varphi_R|^2\,dx\,dy\\
&\quad+d_s\left\{\int_{B_R^+(l,0)}y^a|\nabla u|^2\,dx\,dy\right\}^{\frac{1}{2}}\left\{\int_{B_R^+}y^a|\nabla\varphi_R|^2\,dx\,dy\right\}^{\frac{1}{2}}\\
&\quad+\int_{\partial^0(B_R^+\setminus B_{(1-\eta)R}^+)}b(x)G(w)\,dx\\
&\leq d_s\int_{B_R^+(l,0)}\frac{y^a}{2}|\nabla u|^2\,dx\,dy+\left(C\eta^{-2}R^{-2}RR^{1+a}\right)\\
&\quad+\left\{CR\left[\int_1^R y^a y^{-2}\,dy+\int_0^1\left(y^a+y^{-a}\right)dy\right]\right\}^{\frac{1}{2}}\left(C\eta^{-2}R^a\right)^{\frac{1}{2}}\\
&\quad+2\overline{b}\max_{[0,1]}G\cdot\eta R\\
&\leq d_s\int_{B_R^+(l,0)}\frac{y^a}{2}|\nabla u|^2\,dx\,dy+C\eta^{-2}R^a+C\eta^{-1}R^{\frac{1+a}{2}}+2\overline{b}\max_{[0,1]}G\cdot\eta R.\end{aligned}$$

Here the constant C does not depend on R, we use the gradient estimates (see [2]) in the second line from the bottom.

On the other hand,

$$\mathcal{E}\left(u, B_R^+(l,0)\right) \geq d_s \int_{B_R^+(l,0)} \frac{y^a}{2} |\nabla u|^2 \, dx\, dy + 2\underline{b} \min_{[0,1-\epsilon]} G \cdot R.$$

Choose $\eta = \frac{\underline{b} \min_{[0,1-\epsilon]} G}{2\bar{b} \max_{[0,1]} G}$, $\mathcal{E}(u, B_R^+(l,0)) > \mathcal{E}(w, B_R^+(l,0))$ for large R. This contradiction leads to the result that there exists at least a sequence xn→∞ such that u(xn,0)→1.

Step 3. We show that u is the layer solution, *i.e.*, limx→±∞u(x,0)=±1.

Let un(x,y)=u(x+n,y) and n∈Z+. By the regularity results [2], up to a subsequence,

$$\begin{aligned} &u^n \to u^\infty \quad \text{in } C^0_{\text{loc}}\left(\overline{\mathbb{R}^2_+}\right), \\ &u^n_x \to u^\infty_x \quad \text{in } C^0_{\text{loc}}\left(\overline{\mathbb{R}^2_+}\right), \\ &y^a \frac{\partial u^n}{\partial y} \to y^a \frac{\partial u^\infty}{\partial y} \quad \text{in } C^0_{\text{loc}}\left(\overline{\mathbb{R}^2_+}\right) \end{aligned}$$

as $n \to \infty$.

$$\begin{cases} -\operatorname{div}(y^a \nabla u^\infty) = 0 & \text{in } \mathbb{R}^2_+, \\ -\lim_{y \downarrow 0^+} y^a \frac{\partial u^\infty}{\partial y} = \frac{1}{d_s} b(x) f(u^\infty(x,0)) & \text{on } \partial\mathbb{R}^2_+, \\ 0 \leq u^\infty \leq 1 & \text{in } \mathbb{R}^2_+. \end{cases} \tag{3.11}$$

Define u˜=1−u∞, we have

$$\begin{cases} -\operatorname{div}(y^a \nabla \tilde{u}) = 0 & \text{in } \mathbb{R}^2_+, \\ -\lim_{y \downarrow 0^+} y^a \frac{\partial \tilde{u}}{\partial y} = -\frac{1}{d_s} b(x) f(u^\infty(x,0)) \leq 0 & \text{on } \partial\mathbb{R}^2_+, \\ 0 \leq \tilde{u} \leq 1 & \text{in } \mathbb{R}^2_+. \end{cases} \tag{3.12}$$

$\tilde{u} \equiv C$ by Proposition 3.1, $f(u_\infty(x,0))=f(C)\equiv 0$ and $u_\infty \equiv 0$ or 1. Thus $u_\infty \equiv 1$ by step 2. That is, $u \to 1$ as $x \to \infty$. $u \to -1$ as $x \to -\infty$ is achieved by odd symmetry.

u is the desired layer solution.

Proof of Theorem 1.1 It follows from Theorem 3.1; for the regularity of *v* see [2].

Lastly we give asymptotic estimates for layer solutions of (1.1) as $|x| \to \infty$.

Proof of Theorem 1.2 Let *v* be a layer solution of (1.1),

$$\begin{cases} (-\partial_{xx})^s v(x) = b(x) f(v(x)) & \text{in } \mathbb{R}, \\ \lim_{x\to\pm\infty} v = \pm 1. \end{cases} \tag{3.13}$$

Then

$$(-\partial_{xx})^s (1-v) - b(x) f'(\xi_1)(1-v) = 0 \quad \text{in } \mathbb{R}, \tag{3.14}$$

where ξ_1 is some point between $v(x)$ and 1.

Consider the layer solution v_s^t of the unperturbed problem in Lemma 2.4,

$$(-\partial_{xx})^s \left(1 - v_s^t\right) - \left(f_s^t\right)'(\xi_2)\left(1 - v_s^t\right) = 0 \quad \text{in } \mathbb{R} \tag{3.15}$$

with ξ_2 is some point between $v_s^t(x)$ and 1.

Since $-(f_s^t)'(1) = \frac{1}{t}$, choose t large enough such that $2t < -b-f'(1)$ and choose $x_0 \in \mathbb{R}$ such that $-(f_s^t)'(\xi_2) < \frac{2}{t} < -bf'(\xi_1)$ for all $x > x_0$.

Choose $C>0$ such that $C(1-v_s^t) > 1-v$ in $(-\infty, x_0]$, which can be done since $v_s^t, v \to -1$ as $x \to -\infty$.

Define

$$d(x)=\begin{cases}\frac{2}{t} & \text{in } (x_0,+\infty),\\ \frac{Cf_s^t(v_s^t)-b(x)f(v)}{C(1-v_s^t)-(1-v)} & \text{in } (-\infty,x_0],\end{cases}$$

d(x)∈L∞. We have

$$\begin{cases}(-\partial_{xx})^s\{C(1-v_s^t)-(1-v)\}+d(x)\{C(1-v_s^t)-(1-v)\}\geq 0 & \text{in } \mathbb{R},\\ C(1-v_s^t)-(1-v)>0 & \text{in } (-\infty,x_0].\end{cases} \quad (3.16)$$

Obviously, if infR{C(1−vts)−(1−v)}<0, it is achieved at some point x−∈(x0,+∞). Since $d>0 \text{ in } (x_0,+\infty), (-\partial_{xx})^s\{C(1-v_s^t)-(1-v)\}(\underline{x})\geq 0$ from the first inequality of (3.16), which contradicts with the fact that

$$(-\partial_{xx})^s\left\{C\left(1-v_s^t\right)-(1-v)\right\}(\underline{x})$$
$$=\int_R \frac{\{C(1-v_s^t)-(1-v)\}(\underline{x})-\{C(1-v_s^t)-(1-v)\}(y)}{|\underline{x}-y|^{1+2s}}\,dy<0.$$

Therefore $(1-v)\leq C(1-v_s^t) \text{ for } C>0$ given from above.

On the other hand, choose small t>0 such that −b¯f′(1)<12t and choose x0∈R such that $-\bar{b}f'(\xi_1)<\frac{1}{2t}<-(f_s^t)'(\xi_2)$ Choose c>0 such that $c(1-v_s^t)<1-v \text{ in } (-\infty,x^0]$.

Define

$$\tilde{d}(x)=\begin{cases}\frac{1}{2t} & \text{in } (x^0,+\infty),\\ \frac{b(x)f(v)-cf_s^t(v_s^t)}{(1-v)-c(1-v_s^t)} & \text{in } (-\infty,x^0]\end{cases}$$

and obviously d˜(x)∈L∞.

Then,

$$\begin{cases} (-\partial_{xx})^s\{(1-v)-c(1-v_s^t)\}+\widetilde{d}(x)\{(1-v)-c(1-v_s^t)\}\geq 0 & \text{in } \mathbb{R}, \\ (1-v)-c(1-v_s^t)>0 & \text{in } (-\infty,x^0]. \end{cases} \tag{3.17}$$

If $\inf_R\{(1-v)-c(1-v_s^t)\}<0,$ it is only achieved at some point x¯∈(x0,+∞). Since d˜>0 in $(x^0,+\infty),\ (-\partial_{xx})^s\{(1-v)-c(1-v_s^t)\}(\overline{x})\geq 0$ from the first inequality of (3.17), which contradicts the fact that $(-\partial_{xx})^s\{(1-v)-c(1-v_s^t)\}(\overline{x})<0.$ Thus $c(1-v_s^t)\leq(1-v)$ for some 0<c<C given from above.

Therefore,

$$cx^{-2s}\leq|1-v|\leq Cx^{-2s}\quad \text{for } x>1$$

by Lemma 2.4. Similarly,

$$c|x|^{-2s}\leq|1+v|\leq C|x|^{-2s}\quad \text{for } x<-1.$$

Here *c* and *C* maybe different from above. □

4 ASYMPTOTIC AS *s↑1*

In this section we prove Theorem 1.3, which consists of two lemmas.

Lemma 4.1 *Let* {vsk} *be a sequence of layer solutions of* (1.1) *in Theorem* 1.1. *Then there exists a subsequence denoted again by* {vsk}, *converging locally uniformly to* v1 *which solves the local elliptic equation*

$$-v_{xx}^1(x)=b(x)f\left(v^1\right)\quad in\ \mathbb{R}. \tag{4.1}$$

Proof Consider uak, the *s*-extension of vsk, which solves

$$\begin{cases} -\operatorname{div}(y^{a_k}\nabla u_{a_k}) = 0 & \text{in } \mathbb{R}^2_+, \\ -(1+a_k)\lim_{y\downarrow 0^+} y^{a_k}\partial_y u_{a_k} = C_{a_k} b(x) f(u_{a_k}(x,0)) & \text{on } \partial\mathbb{R}^2_+, \end{cases} \tag{4.2}$$

where $a_k = 1 - 2s_k$ and $C_{a_k} = \frac{1+a_k}{d_{s_k}} = \frac{2(1-s_k)}{d_{s_k}}$. Obviously ak↓−1 as sk↑1. Let $\xi \in C_c^1(\overline{\mathbb{R}^2_+})$. Multiplying (4.2) with ξ and integrating in R2+,

$$(1+a_k)\int_{\mathbb{R}^2_+} y^{a_k}\nabla u_{a_k}\nabla\xi \, dx\, dy - C_{a_k}\int_{\mathbb{R}} b(x) f\left(u_{a_k}(x,0)\right)\xi \, dx = 0. \tag{4.3}$$

Choose $\xi(x,y) = \xi_1(x)\xi_2(y)$, $\xi_1 \in C_c^1(\mathbb{R})$ and $\xi_2$is a cut-off function which equals 1 in [0,1] and 0 in [2,∞), |ξ′2|≤C for some constant C>0. Thus (4.3) can be rewritten as

$$\begin{aligned}(1+a_k)\int_{\mathbb{R}^2_+} y^{a_k}\left\{\xi_1'(x)\xi_2(y)\partial_x u_{a_k} + \xi_1(x)\xi_2'(y)\partial_y u_{a_k}\right\} dx\, dy \\ = C_{a_k}\int_{\mathbb{R}} b(x) f\left(u_{a_k}(x,0)\right)\xi_1(x)\, dx.\end{aligned} \tag{4.4}$$

By the regularity results in [2], the continuous module does not depend on *s* for s>s0>12. Up to a subsequence,

$$\begin{aligned} &u_{a_k} \to u_{-1} \quad \text{in } C^0_{\text{loc}}\left(\overline{\mathbb{R}^2_+}\right), \\ &(u_{a_k})_x \to (u_{-1})_x \quad \text{in } C^0_{\text{loc}}\left(\overline{\mathbb{R}^2_+}\right) \end{aligned}$$

and

$$C_{a_k} = \frac{2(1-s_k)}{d_{s_k}} = \frac{2(1-s_k)}{2^{2s_k-1}\frac{\Gamma(s_k)}{\Gamma(1-s_k)}} \to 1$$

as sk↑1 (or equivalently ak↓−1). Then

$$C_{a_k}\int_{\mathbb{R}} b(x) f\left(u_{a_k}(x,0)\right)\xi_1\, dx \to \int_{\mathbb{R}} b(x) f\left(u_{-1}(x,0)\right)\xi_1\, dx \quad \text{as } a_k \downarrow -1. \tag{4.5}$$

For the first integral in (4.4), we consider

$$\begin{aligned}
&(1+a_k)\int_0^\infty y^{a_k}\xi_2(y)\partial_x u_{a_k}\,dy \\
&\quad = (1+a_k)\int_0^\delta y^{a_k}\xi_2(y)\left\{\partial_x u_{a_k} - u'_{-1}(x)\right\}dy \\
&\qquad + (1+a_k)\int_0^\delta y^{a_k}\xi_2(y)u'_{-1}(x)\,dy + (1+a_k)\int_\delta^\infty y^{a_k}\xi_2(y)\partial_x u_{a_k}\,dy \\
&\quad = I_1 + I_2 + I_3,
\end{aligned} \tag{4.6}$$

$$|I_1| \le (1+a_k)\int_0^\delta y^{a_k}\xi_2(y)\left|\partial_x u_{a_k} - u'_{-1}(x)\right|dy \le \epsilon\delta^{1+a_k} \tag{4.7}$$

for 0<δ<1 and small ϵ>0. Here we use the fact that $\partial_x u_{a_k} \to u'_{-1}(x)$ locally uniformly in $\overline{\mathbb{R}^2_+}$. We have

$$I_2 = u'_{-1}(x)(1+a_k)\int_0^\delta y^{a_k}\,dy = \delta^{1+a_k}u'_{-1} \to u'_{-1} \quad \text{as } a_k \downarrow -1. \tag{4.8}$$

Since $|\nabla u_{a_k}| \le \frac{C}{y}$ for $y>0$ and C independent of ak (see [2]),

$$|I_3| \le C(1+a_k)\int_\delta^\infty y^{a_k-1}\,dy = C\frac{1+a_k}{a_k}\delta^{a_k} \to 0 \quad \text{as } a_k \downarrow -1. \tag{4.9}$$

By (4.6)-(4.9),

$$(1+a_k)\int_0^\infty y^{a_k}\xi_2(y)\partial_x u_{a_k}\,dy \to u'_{-1}$$

and

$$(1+a_k)\int_{\mathbb{R}^2_+} y^{a_k}\xi'_1(x)\xi_2(y)\partial_x u_{a_k}\,dx\,dy \to \int_{\mathbb{R}}\xi'_1(x)u'_{-1}\,dx, \tag{4.10}$$

$$\left|(1+a_k)\int_{\mathbb{R}_+^2} y^{a_k}\xi_1(x)\xi_2'(y)\partial_y u_{a_k}\,dx\,dy\right|$$
$$\le \int_{\mathbb{R}}|\xi_1(x)|\,dx(1+a_k)\int_1^2 y^{a_k}|\xi_2'(y)||\partial_y u_{a_k}|\,dy$$
$$\le C(1+a_k)\int_1^2 y^{a_k-1}\,dy = C\frac{1+a_k}{a_k}(2^{a_k}-1)\to 0 \tag{4.11}$$

as ak↓−1.

Therefore, by (4.4), (4.5), (4.10), and (4.11),

$$\int_{\mathbb{R}} u_{-1}'(x)\xi_1'(x)\,dx = \int_{\mathbb{R}} b(x)f(u_{-1}(x))\xi_1(x)\,dx. \tag{4.12}$$

That is,

$$-v_{xx}^1 = b(x)f(v^1) \tag{4.13}$$

in the weak sense (u−1=v1). By the regularity theory of elliptic equations, v1 is also a classical solution of (4.13). □

Lemma 4.2 v1 *is also a layer solution of* (4.1), *i.e.*, v1→±1 *as* x±∞.

Proof Claim 1. v1 is a local minimizer in (0,∞) under perturbations in [−1,1]. That is,

$$\mathcal{F}(v^1,I)\le\mathcal{F}(v^1+\xi_1,I) \tag{4.14}$$

for any bounded open interval I⊂(0,∞) and for any ξ1∈C10(I) such that |v1+ξ1|≤1, where

$$\mathcal{F}(w,I) := \int_I\left\{\frac{|w_x|^2}{2}+b(x)G(w)\right\}dx \quad \text{for every } w\in H^1(I).$$

Indeed, for the test function ξ in Lemma 4.1 with the additional property that |uak+ξ|≤1, we have

$$
\begin{aligned}
0 &\le E\big(u_{a_k} + (1-\epsilon)\xi, I\times[0,R]\big) - E\big(u_{a_k}, I\times[0,R]\big)\\
&= \frac{1+a_k}{2}\int_{I\times[0,R]} y^{a_k}\left|\nabla\big(u_{a_k} + (1-\epsilon)\xi\big)\right|^2 dx\,dy + C_{a_k}\int_I b(x)G\big(u_{a_k} + (1-\epsilon)\xi\big)\,dx\\
&\quad - \frac{1+a_k}{2}\int_{I\times[0,R]} y^{a_k}|\nabla u_{a_k}|^2\,dx\,dy - C_{a_k}\int_I b(x)G(u_{a_k})\,dx\\
&= \frac{1+a_k}{2}\int_{I\times[0,R]} y^{a_k}\left|\partial_x u_{a_k} + (1-\epsilon)\xi_1'(x)\xi_2(y)\right|^2 dx\,dy\\
&\quad - \frac{1+a_k}{2}\int_{I\times[0,R]} y^{a_k}|\partial_x u_{a_k}|^2\,dx\,dy\\
&\quad + (1+a_k)\int_{I\times[0,R]} y^{a_k}\partial_y u_{a_k}(1-\epsilon)\xi_1(x)\xi_2'(y)\,dx\,dy\\
&\quad + \frac{1+a_k}{2}\int_{I\times[0,R]} y^{a_k}\big((1-\epsilon)\xi_1(x)\xi_2'(y)\big)^2 dx\,dy\\
&\quad + C_{a_k}\int_I b(x)G\big(u_{a_k} + (1-\epsilon)\xi_1(x)\big)\,dx - C_{a_k}\int_I b(x)G(u_{a_k})\,dx.
\end{aligned}
\tag{4.15}
$$

As in the discussions in Lemma 4.1, let ak↓−1, and we have

$$
\frac{1+a_k}{2}\int_{I\times[0,R]} y^{a_k}(\partial_x u_{a_k})^2\,dx\,dy \to \int_I \frac{(u_{-1}')^2}{2}\,dx,
\tag{4.16}
$$

$$
(1+a_k)\int_{I\times[0,R]} y^{a_k}\partial_x u_{a_k}(1-\epsilon)\xi_1'(x)\xi_2(y)\,dx\,dy \to \int_I u_{-1}'(x)(1-\epsilon)\xi_1'(x)\,dx,
\tag{4.17}
$$

$$
\begin{aligned}
&\frac{1+a_k}{2}\int_{I\times[0,R]} y^{a_k}\big((1-\epsilon)\xi_1'(x)\xi_2(y)\big)^2 dx\,dy\\
&= \frac{1}{2}\int_I (1-\epsilon)^2\big(\xi_1'(x)\big)^2 dx\left\{\int_0^1 (1+a_k)y^{a_k}\,dy + \int_1^R (1+a_k)y^{a_k}\big(\xi_2(y)\big)^2 dy\right\}\\
&\to \frac{1}{2}\int_I \big((1-\epsilon)\xi_1'(x)\big)^2 dx.
\end{aligned}
\tag{4.18}
$$

By (4.16)-(4.18),

$$\frac{1+a_k}{2}\int_{I\times[0,R]} y^{a_k}\left(\partial_x u_{a_k} + (1-\epsilon)\xi_1'(x)\xi_2(y)\right)^2 dx\,dy - \frac{1+a_k}{2}\int_{I\times[0,R]} y^{a_k}(\partial_x u_{a_k})^2\,dx\,dy \to \frac{1}{2}\int_I \left(u_{-1}'(x) + (1-\epsilon)\xi_1'(x)\right)^2 dx - \frac{1}{2}\int_I \left(u_{-1}'(x)\right)^2 dx, \tag{4.19}$$

$$\begin{aligned}(1+a_k)\int_{I\times[0,R]} y^{a_k}\partial_y u_{a_k}\xi_1(x)\xi_2'(y)\,dx\,dy &= (1+a_k)\int_{I\times[1,2]} y^{a_k}\partial_y u_{a_k}\xi_1(x)\xi_2'(y)\,dx\,dy \\ &\le C(1+a_k)\int_1^2 y^{a_k-1}\,dy \\ &= \frac{C(1+a_k)}{a_k}\{2^{a_k}-1\} \to 0 \quad \text{as } a_k \downarrow -1,\end{aligned} \tag{4.20}$$

$$\begin{aligned}\frac{(1+a_k)}{2}\int_I\int_0^R y^{a_k}\left(\xi_1(x)\xi_2'(y)\right)^2 dx\,dy &= \frac{(1+a_k)}{2}\int_I\int_1^2 y^{a_k}\left(\xi_1(x)\xi_2'(y)\right)^2 dx\,dy \\ &\le C(1+a_k)\int_1^2 y^{a_k}\,dy \\ &= C(2^{a_k+1}-1) \to 0 \quad \text{as } a_k \downarrow -1,\end{aligned} \tag{4.21}$$

$$C_{a_k}\int_I b(x)G\left(u_{a_k} + (1-\epsilon)\xi_1(x)\right)dx - C_{a_k}\int_I b(x)G(u_{a_k})\,dx \to \int_I b(x)G\left(u_{-1} + (1-\epsilon)\xi_1(x)\right)dx - \int_I b(x)G(u_{-1})\,dx. \tag{4.22}$$

By (4.15), (4.19)-(4.22), our claim is proved.

Claim 2. $v^1 \to 1$ as $x \to \infty$.

Define $v^{1,n}(x) = v^1(x+n)$ for $n \in \mathbb{Z}^+$, up to a subsequence, $v^{1,n} \to v^{1,\infty}$ in C^2_{loc} as $n \to \infty$,

$$\begin{cases} -v_{xx}^{1,\infty}(x) = b(x)f(v^{1,\infty}(x)), & x \in \mathbb{R}, \\ 0 \le v^{1,\infty} \le 1. \end{cases} \tag{4.23}$$

Since $f \ge 0$ and $b > 0$, $-v_{xx}^{1,\infty} \ge 0$ in $\mathbb{R}$ and $v^{1,\infty} \equiv 0$ or $1.$.

We show that $v^1 \to 0$ or 1 as $x \to \infty.$. Indeed, if there are two sequences {xn} and {yn} such that

$v^1(x_n) \to 0$ and $v^1(y_n) \to 1$ as $n \to \infty$, there must exist $z_n \in (x_n, y_n)$ such that $v^1(z_n) = \frac{1}{2}.$.

Denote $\widetilde{v}_n^1(x) = v^1(x + [z_n])$ where $[z_n]$ is the integer part of $\widetilde{v}_n^1(z_n - [z_n]) = v^1(z_n) = \frac{1}{2}$ and up to a subsequence $\widetilde{v}_n^1 \to \widetilde{v}_\infty^1$ in C_{loc}^2, $\widetilde{v}_\infty^1$ solves equation (4.23). Therefore v˜1∞≡0 or 1. For the above subsequence, there is a subsubsequence such that zn−[zn]→z∗∈[0,1] as n→∞ and v˜1∞(z∗)=12. This contradiction verifies v1→0 or 1 as x→∞.

To check v1→1 as x→∞, suppose that v1→0 as x→∞ by contradiction. Then,

$$\liminf_{l \to +\infty} \mathcal{F}\left(v^1, (l-R, l+R)\right) = \liminf_{l \to +\infty} \int_{l-R}^{l+R} \left\{ \frac{|v^1|^2}{2} + b(x)G\left(v^1\right) \right\} dx \ge 2\underline{b}R\epsilon$$

for some ϵ>0.

Let ξ∈C10(l−R,l+R), ξ=1 if |x−l|<(1−η)R and ξ=0 if |x−l|>R where η will be determined later, |ξ′|≤1ηR. Define w=1⋅ξ+(1−ξ)v1, then w(l±R)=v1(l±R). We have

$$\limsup_{l\to+\infty} \mathcal{F}(w,(l-R,l+R))$$

$$= \limsup_{l\to+\infty} \int_{l-R}^{l+R} \left(\frac{1}{2} \left|(1-\xi)v_x^1 + (1-v^1)\xi_x\right|^2 + b(x)G(1\cdot\xi + (1-\xi)v^1) \right) dx$$

$$\leq \int_{l-R}^{l+R} \xi_x^2\, dx + \overline{b} \max_{[-1,1]} G \cdot 2\eta R$$

$$\leq \frac{1}{\eta^2 R} + \overline{b} \max_{[-1,1]} G \cdot 2\eta R.$$

Choose $\eta = \frac{\epsilon \underline{b}}{2\overline{b} \max_{[-1,1]} G}$,

$$\limsup_{l\to+\infty} \mathcal{F}(w,(l-R,l+R)) < \liminf_{l\to+\infty} \mathcal{F}(v^1,(l-R,l+R))$$

for R>1 large enough. Therefore v1→1 as x→∞, by odd symmetry, v1→−1 as x→−∞, *i.e.*, v1 is a layer solution of the local elliptic equation (4.13).

By the Hamiltonian equality (2.15),

$$b(x)\{G(v^1(x)) - G(1)\} + \int_x^{+\infty} b'(t)\{G(v^1(t)) - G(1)\}\, dt$$

$$= \frac{1}{2}(v_x^1)^2 = \lim_{a_k\downarrow -1} (1+a_k) \int_0^\infty \frac{y^{a_k}}{2} (\partial_x u_{a_k})^2$$

$$= \lim_{a_k\downarrow -1} (1+a_k) \int_0^\infty \frac{y^{a_k}}{2} (\partial_y u_{a_k})^2$$

$$+ \lim_{a_k\downarrow -1} C_{a_k} b(x)\{G(u_{a_k}(x,0)) - G(1)\}$$

$$+ \lim_{a_k\downarrow -1} C_{a_k} \int_x^\infty b'(t)\{G(u_{a_k}(t,0)) - G(1)\}\, dt.$$

Therefore,

$$\lim_{a\downarrow -1}(1+a) \int_0^\infty \frac{y^a}{2} (\partial_y u_a)^2 = \int_x^{+\infty} b'(t)\{G(v^1(t)) - G(1)\}\, dt$$

$$- \lim_{a_k\downarrow -1} C_{a_k} \int_x^\infty b'(t)\{G(u_{a_k}(t,0)) - G(1)\}\, dt.$$

□

Proof of Theorem 1.3 It follows from Lemmas 4.1 and 4.2. □

ACKNOWLEDGEMENTS

This research has been supported by National Natural Science Foundation of China (Grant No. 11371128).

COMPETING INTERESTS

The author declares that she has no competing interests.

AUTHOR'S CONTRIBUTIONS

The author read and approved the final manuscript.

REFERENCES

1. Cabré X, Solà-Morales J: Layer solutions in a half-space for boundary reactions. *Commun. Pure Appl. Math.* 2005, 58(12): 1678-1732.

2. Cabré X, Sire Y: Nonlinear equations for fractional Laplacians I: regularity, maximum principles, and Hamiltonian estimates. *Ann. Inst. Henri Poincaré, Anal. Non Linéaire* 2013. 10.1016/ j.anihpc.2013.02.001

3. Cabré, X, Sire, Y: Nonlinear equations for fractional Laplacian II: existence, uniqueness, and qualitative properties of solutions. arXiv:1111.0796 (2011)

4. Caffarelli L, Silvestre L: An extension problem related to the fractional Laplacian. *Commun. Partial Differ. Equ.* 2007, 32(8):1245-1260.

5. Fabes EB, Kenig CE, Serapioni RP: The local regularity of solutions of degenerate elliptic equations. *Commun. Stat., Theory Methods*1982, 7(1):77-116.

6. Palatucci G, Savin O, Valdinoci E: Local and global minimizers for a variational energy involving a fractional norm. *Ann. Mat. Pura Appl.*2013, 192(4):673-718.

7. Nekvinda A: Characterization of traces of the weighted Sobolev space W1,p(Ω,dεM) on *M* . *Czechoslov. Math. J.* 1993, 43(4):695-711.

8. Alama S, Bronsard L, Gui C:Stationary layered solutions in R2 for an Allen-Cahn systems with multiple well potential. *Calc. Var. Partial Differ. Equ.* 1997, 5(4):359-390.

9. Ghoussoub N, Gui C: On a conjecture of De Giorgi and some related problems. *Math. Ann.* 1998, 311: 481-491.

10. Ghoussoub N, Gui C: On De Giorgi's conjecture in dimensions 4 and 5. *Ann. Math.* 2003, 157(1):313-334.

11. Adams RA Pure and Applied Mathematics 65. In *Sobolev Spaces.* Academic Press, New York; 1975.

12. Gilbarg D, Trudinger NS: *Elliptic Partial Differential Equations of Second Order.* Springer, Berlin; 1997.

7

EXISTENCE OF SOLUTIONS FOR SEMILINEAR ELLIPTIC EQUATIONS ON R*N*

Ruichang Pei1,2* and Jihui Zhang1

1Institute of Mathematics, School of Mathematics and Computer Sciences, Nanjing Normal University

2School of Mathematics and Statistics, Tianshui Normal University

ABSTRACT

In this paper, the existence of at least one nontrivial solution for a class of semilinear elliptic equations on $\mathbb{R}^N$ is established by using the linking methods.

KEYWORDS

Schrödinger equation, subcritical exponent, local linking

1 INTRODUCTION

In this paper we consider the question of the existence of solutions for a class of semilinear equations of the form

$$(P_\lambda) \quad -\Delta u + \lambda u = g(x,u), \quad x \in \mathbb{R}^N,$$

where $\lambda>0$ is a parameter and the nonlinearity $g \in C(\mathbb{R}^N \times \mathbb{R})$ is asymptotically linear, *i.e.*,

$$\lim_{|t|\to\infty} \frac{g(x,t)}{t} = V(x), \qquad \lim_{|x|\to\infty} V(x) = v_\infty \tag{1.1}$$

for some $V(x) \in C(\mathbb{R}^N, \mathbb{R})$ and $v_\infty \in \mathbb{R}$. In case this equation is considered in a bounded domain $\Omega \subset \mathbb{R}^N$ (with, say, the Dirichlet boundary condition), there is a large amount of literature on existence and multiplicity results, with the case of resonance being of particular interest (see [1–3]). We recall that the problem is said to be at resonance if $-\lambda\in\sigma(S)$$-\lambda \in \sigma(S)$, where $\sigma(S)$$\sigma(S)$ denotes the spectrum of S, the 'asymptotic linearization' of the problem. In other words, $S : D(S) \subset L^2(\Omega) \to L^2(\Omega)$ is the operator given by

$$Su(x) = -\Delta u(x) - V(x)u(x), \qquad D(S) = H_0^1(\Omega) \cap H^2(\Omega). \tag{1.2}$$

On the other hand, a systematic study of such asymptotically linear problems set in unbounded domains or the whole space $\mathbb{R}^N$ is more recent and presents a number of mathematical difficulties (see [4, 5]). As an example, we note that in the case of problem ($P\lambda$$P_\lambda$), the asymptotic linearization operator S (now defined on $D(S) = H^2(\mathbb{R}^N)$) has a much more complicated spectrum (including an *essential* part $[-v\infty,\infty)$$[-v_\infty, \infty)$), which in turn makes the study of this problem more challenging. In [4], motivated by the paper [5], Tehrani and Costa studied the existence of positive solutions to ($P\lambda$) by using the mountain pass theorem if $g(x,u)$ satisfies some strong asymptotically linear conditions. Comparing with previous paper [4], in [6], Tehrani obtained the existence of a (possibly sign-changing) solution for problem ($P\lambda$) under essentially condition (1.1) only. In fact, he proved the following.

Theorem 1.0 [6] *Let* $g0(x,s):=g(x,s)-V(x)s$ *and assume that*

(G)for every ϵ>0$\epsilon > 0$, *there exists* $0 \le b_\epsilon(x) \in L^2(\mathbb{R}^N)$ *such that*

$$|g_0(x,s)| \le b_\epsilon(x) + \epsilon|s| \quad a.e.\ x \in \mathbb{R}^N, s \in \mathbb{R}.$$

If Λ≥0 *or* max{0,v∞}<λ<−Λ *and* −λ∉σp(S), *then* (Pλ) *has a solution in* H1(RN).

Now, one naturally asks: Are there nontrivial solutions for problem (Pλ) if −λ∈σ(S)$-\lambda \in \sigma(S)$ in the above theorem? Obviously, this case is resonance. But, this problem is not easy because we face the difficulties of verifying that the energy functional satisfies the (PS) condition if we still follow the idea of [6]. Here, there is still an interesting problem: Are there nontrivial solutions for problem (PλP_λ) if−λ∈σ(S) and g0(x,s) (in Theorem 1.0) is more generalized superlinear? We will answer the above problems affirmatively by using Li and Willem's local linking methods (see [7]).

Next, we recall a few basic facts in the theory of Schrödinger operators which are relevant to our discussion (see [6]).

*1.*Since $\lim_{|x|\to\infty} V(x) = v_\infty$, one has $\sigma_{ess}(S) = [-v_\infty, \infty)$.

*2.*The bottom of the spectrum σ(S) of the operator *S* is given by

$$\Lambda = \lambda_0 = \inf_{0 \ne u \in H^2(\mathbb{R}^N)} \frac{\int |\nabla u|^2 - V(x)u^2}{\int u^2}.$$

Therefore we clearly have Λ≤−v∞. If Λ<−v∞, then by using the concentration compactness principle of Lions, one shows that Λ is the principle eigenvalue of *S* with a positive eigenfunction Φ0:

$$S\Phi_0 = \lambda_0\Phi_0, \quad \Phi_0 \in H^2(\mathbb{R}^N), \Phi_0 > 0.$$

*3.*The spectrum of *S* in (−∞,−v∞), namely σ(S)∩(−∞,−v∞), is at most a countable set, which we denote by

$\Lambda=\lambda_0<\lambda_1<\lambda_2<\cdots,$

where each λ_k is an isolated eigenvalue of S of the finite multiplicity. Let E_{λ_j} denote the eigenspace of S corresponding to the eigenvalue λ_j.

Now, we state our main results. In this paper, we always assume that $\lim_{|x|\to\infty}V(x)=v_\infty$ and $v_\infty<0$. The conditions imposed on $g_0(x,t)$ (see Theorem 1.0) are as follows:

(H_1) $g_0\in C(\mathbb{R}^N\times\mathbb{R},\mathbb{R})$, and there are constants $C_1,C_2\geq 0$ such that

$$|g_0(x,t)| \leq C_1 + C_2|t|^{s-1}, \quad \forall x \in \mathbb{R}^N, \forall t \in R, s \in (2,p^*)\ (N \geq 3),$$

where $p^* = \frac{2N}{N-2}$;

(H_2) $g_0(x,t) = o(|t|),\ |t| \to 0$, uniformly on $\mathbb{R}^N$;;

(H_3) $\lim_{|t|\to\infty} \frac{g_0(x,t)}{t} = +\infty$ uniformly on $\mathbb{R}^N$;

(H_4) There is a constant $\theta\geq 1$ such that for all $(x,t)\in\mathbb{R}^N\times\mathbb{R}$ and $s\in[0,1]$,

$$\theta\left(g_0(x,t)t - 2G_0(x,t)\right) \geq \left(sg_0(x,st)t - 2G_0(x,st)\right),$$

where $G_0(x,t) = \int_0^t g_0(x,s)\,ds$;

(H_5) For some $\delta>0$, either

$$G_0(x,t) \geq 0 \quad \text{for } |t| \leq \delta, x \in \mathbb{R}^N$$

or

$$G_0(x,t) \leq 0 \quad \text{for } |t| \leq \delta, x \in \mathbb{R}^N;$$

(H$_6$) $\lim_{|x|\to\infty} \sup_{|t|\le r} \frac{g_0(x,t)}{|t|} = 0$ for every $r > 0$.

Theorem 1.1 *Assume that conditions* (H$_1$)-(H$_4$) *hold. If* $-\lambda$ *is an eigenvalue of* S($-\lambda<-v\infty$), *assume also that* (H$_5$) *and* (H$_6$)*hold. Then the problem* (Pλ P_λ) *has at least one nontrivial solution.*

Remark 1.1 It follows from the condition (H$_3$) that our nonlinearity g0(x,t) does not satisfy the classical condition of Ambrosetti and Rabinowitz:

(AR) There is $\mu>2$ such that $0<\mu G0(x,u)\le ug0(x,u)$ for all $x\in RN$ and $u\neq 0$ $u \neq 0$.

In recent years, there have been some papers devoted to replacing (AR) with more natural conditions (see [8–10]). But our methods are different from the references therein.

We also consider asymptotically quadratic functions. We assume that:

(H$_7$) For every $\epsilon>0$ $\epsilon > 0$, there exists $0\le b\epsilon(x)\in L^2(R^N)$ such that

$$|g_0(x,s)| \le b_\epsilon(x) + \epsilon|s| \quad \text{a.e. } x \in \mathbb{R}^N, s \in R,$$

and $\lambda_k < -\lambda < \lambda_{k+1}$.

Theorem 1.2 *Assume that conditions* (H$_2$), (H$_6$), (H$_7$) *and one of the following conditions hold*:

(A_1) $\lambda_j < 0 < \lambda_{j+1}, j \neq k$;
(A_2) $\lambda_j = 0 < \lambda_{j+1}, j \neq k$ *for some* $\delta > 0$,

$$G_0(x,u) \geq 0 \quad for\ |u| > \delta, x \in \mathbb{R}^N;$$

(A_3) $\lambda_j < 0 = \lambda_{j+1}, j \neq k$ *for some* $\delta > 0$,

$$G_0(x,u) \geq 0 \quad for\ |u| \leq \delta, x \in \mathbb{R}^N.$$

Then problem (PλP_λ) *has at least one nontrivial solution.*

2 PRELIMINARIES

Let X be a Banach space with a direct sum decomposition

$X=X^1 \oplus X^2$.

Consider two sequences of subspaces

$$X_0^1 \subset X_1^1 \subset \cdots \subset X^1, \qquad X_0^2 \subset X_1^2 \subset \cdots \subset X^2$$

such that

$$X^j = \bigcup_{n \in N} X_n^j, \quad j = 1,2.$$

For every multi-index $\alpha = (\alpha_1, \alpha_2) \in N^2$, let $X_\alpha = X_{\alpha_1} \oplus X_{\alpha_2}$. We know that

$$\alpha \leq \beta \quad \Leftrightarrow \quad \alpha_1 \leq \beta_1, \qquad \alpha_2 \leq \beta_2.$$

A sequence $(\alpha_n) \subset N^2$ is admissible if, for every $\alpha \in N^2$, there is $m \in N$ such that $n \geq m \Rightarrow \alpha_n \geq \alpha$. For every $I: X \to R$, we denote by $I\alpha I_\alpha$ the function I restricted $X\alpha X_\alpha$.

Definition 2.1 Let *I* be locally Lipschitz on *X* and c∈R. The functional *I* satisfies the (C)∗c condition if every sequence (uαn) such that (αn) is admissible and

$$\|u_1 + u_2\|^2 = \|u_1\|^2 + \|u_2\|^2, \quad u_j \in X_j, j = 1, 2.$$

contains a subsequence which converges to a critical point of *I*.

Definition 2.2 Let *I* be locally Lipschitz on *X* and c∈R$c \in R$. The functional *I* satisfies the (C)∗$(C)^*$ condition if every sequence (uαn)(u_{α_n})such that (αn)(α_n) is admissible and

$$u_{\alpha_n} \in X_{\alpha_n}, \quad I(u_{\alpha_n}) \to c, \qquad (1 + \|u_{\alpha_n}\|)I'(u_{\alpha_n}) \to 0$$

contains a subsequence which converges to a critical point of *I*.

Remark 2.1 1. The (C)∗$(C)^*$ condition implies the (C)∗c$(C)^*_c$ condition for every c∈R$c \in R$.

*2.*When the (C)∗c$(C)^*_c$ sequence is bounded, then the sequence is a (PS)∗c$(PS)^*_c$ sequence (see [11]).

*3.*Without loss of generality, we assume that the norm in *X* satisfies

$$\|u_1 + u_2\|^2 = \|u_1\|^2 + \|u_2\|^2, \quad u_j \in X_j, j = 1, 2.$$

Definition 2.3 Let *X* be a Banach space with a direct sum decomposition

$$X = X_1 \oplus X_2.$$

The function I∈C1(X,R) has a local linking at 0, with respect to (X1,X2) if, for some r>0,

$$I(u) \geq 0, \quad u \in X^1, \|u\| \leq r,$$

$$I(u) \leq 0, \quad u \in X^2, \|u\| \leq r.$$

Lemma 2.1 (see [7])

Suppose that I∈C1(X,R)$I \in C^1(X,R)$ *satisfies the following assumptions*:

(B_1) *I has a local linking at* 0 *and* X1≠{0}$X^1 \neq \{0\}$;

(B_2) *I satisfies* (PS)∗$(PS)^*$;

(B_3) *I maps bounded sets into bounded sets*;

(B_4) *For every* m∈N, $I(u) \to -\infty,\ \|u\| \to \infty,\ u \in X = X_m^1 \oplus X^2$. *Then I has at least two critical points.*

Remark 2.2 Assume that *I* satisfies the $(C)_c^*$ condition. Then this theorem still holds.

Let *X* be a real Hilbert space and let $I \in C^1(X,R)$. The gradient of *I* has the form

$$\nabla I(u) = Au + B(u),$$

where *A* is a bounded self-adjoint operator, 0 is not the essential spectrum of *A*, and *B* is a nonlinear compact mapping.

We assume that there exist an orthogonal decomposition,

$$X = X_1 + X_2,$$

and two sequences of finite-dimensional subspaces,

$$X_0^1 \subset X_1^1 \subset X_1^1 \subset \cdots \subset X^1, \qquad X_0^2 \subset X_1^2 \subset \cdots \subset X^2,$$

such that

$$X^j = \overline{\bigcup_{n \in N} X_n^j}, \quad j = 1,2,$$

$$AX_n^j \subset X_n^j, \quad j = 1,2, n \in N.$$

For every multi-index $\alpha=(\alpha 1,\alpha 2)\in N^2$, we denote by $X\alpha$ the space

$$X_\alpha^1 \oplus X_\alpha^2,$$

by $p\alpha:X\to X\alpha$ the orthogonal projector onto $X\alpha X_\alpha$, and by $M^-(L)$ the Morse index of a self-adjoint operator *L*.

Lemma 2.2 (see [7]) *I satisfies the following assumptions*:

(i)I has a local linking at 0 *with respect to* (X^1,X^2);

(ii)There exists a compact self-adjoint operator $B\infty$ *such that*

$$B(u) = B_\infty(u) + o(\|u\|), \quad \|u\| \to \infty;$$

(iii)$A+B\infty$ *is invertible*;

(vi)For infinitely many multiple-indices $\alpha:=(n,n)$,

$$M^-((A + P_\alpha B_\infty)|_{X_\alpha}) \neq \dim X_n^2.$$

Then I has at least two critical points.

3 THE PROOF OF MAIN RESULTS

Proof of Theorem 1.1 (1) We shall apply Lemma 2.1 to the functional

$$I(u) = \frac{1}{2}\int (|\nabla u|^2 - V(x)|u|^2) + \frac{1}{2}\lambda \int u^2 - \int_\Omega G_0(x,u)$$

defined on $X=H1(RN)$. We consider only the case $-\lambda\in\sigma(S)$, and

$$G_0(x,u) \le 0 \quad \text{for } |u| \le \delta, x \in \mathbb{R}^N. \tag{3.1}$$

Then other case is similar and simple.

Let X2 be a finite dimensional space spanned by the eigenfunctions corresponding to negative eigenvalues of S+λ and let X^1be its orthogonal complement in *X*. Choose a Hilbertian basis ene_n (n≥0) for *X* and define

$$X_n^1 = \operatorname{span}(e_0, e_1, \ldots, e_n), \quad n \in N;$$

$$X_n^2 = X^2, \quad n \in N;$$

$$X^1 = \overline{\bigcup_{n \in N} X_n^1}.$$

By the condition (H_1) and Sobolev inequalities, it is easy to see that the functional *I* belongs to $C^1(X,R)$ and maps bounded sets to bounded sets.

(2) We claim that *I* has a local linking at 0 with respect to (X^1,X^2). Decompose X1 into

$V + W$ when $V = E_{-\lambda}$, $W = (X^2 + V)^{\perp}$. Also, set $u = v + w$, $u \in X^1$, $v \in V$, $w \in W$.

For the convenience of our proof, we state some facts for the norm of the whole space *X*. It is well known that there is an equivalent norm ||·|| on X=$H^1(R^N)$ such that

$$\int (|\nabla u|^2 - V(x)|u|^2) = -\|u\|^2, \quad u \in X^2$$

and

$$\int (|\nabla u|^2 - V(x)|u|^2) = \|u\|^2, \quad u \in W.$$

By the equivalence of norm in the finite-dimensional space, there exists C>0 such that

$$\|v\|_\infty \leq C\|v\|, \quad \forall v \in V. \tag{3.2}$$

It follows from (H_1) and (H_2) that for any $\epsilon>0$, there exists C_ϵ such that

$$|G_0(x,u)| \leq \epsilon u^2 + C_\epsilon |u|^s. \tag{3.3}$$

Hence, we obtain

$$I(u) \leq -m\|u\|^2 + c^*\|u\|^{s+1},$$

where $m>0$, $c*$ is a constant and hence, for $r>0$ small enough,

$$I(u) \leq 0, \quad u \in X^2, \|u\|_X \leq r.$$

Let $u = v + w \in X^1$ be such that $\|u\|_X \leq r_1 = \frac{\delta}{2C}$ and let

$$\mathbb{A}_1 = \left\{x \in \mathbb{R}^N : |w(x)| \leq \frac{\delta}{2}\right\},$$

$$\mathbb{A}_2 = \mathbb{R}^N \setminus \mathbb{A}_1.$$

From (3.2), we have

$$|v(x)| \leq \|v\|_\infty \leq C\|v\| \leq \frac{\delta}{2}$$

for all $\|u\| \leq r_1$ and $x \in \mathbb{R}^N$. On the one hand, one has $|u(x)| \leq |v(x)| + |w(x)| \leq \|v\|_\infty + \frac{\delta}{2} \leq \delta$ for all $x \in A1$. Hence, from (H_5), we obtain

$$\int_{\mathbb{A}_1} G_0(x,u)\,dx \leq 0.$$

On the other hand, we have

$$|u(x)| \le |v(x)| + |w(x)| \le \frac{\delta}{2} + |w(x)| \le 2|w(x)|$$

for all x∈A2. It follows from (3.3) that

$$G_0(x,u) \le \epsilon u^2 + C_\epsilon |u|^{s+1} \le 4\epsilon w^2 + 2^{s+1} C_\epsilon |w|^{s+1}$$

for all x∈A2 and all u∈X1 with ‖u‖≤r1, which implies that

$$\begin{aligned}\int G_0(x,u) &\le 4\epsilon \int_{\mathbb{A}_2} w^2\,dx + \int_{\mathbb{A}_2} 2^{s+1} C_\epsilon |w|^{s+1}\,dx \\ &\le 4(C_3)^2 \epsilon \|w\|^2 + (2C_3)^{\lambda+1} C_\epsilon \|w\|^{s+1},\end{aligned}$$

where C3 is a constant. Hence, there exist positive constants C∗∗, C4 and C5C_5 such that

$$\begin{aligned}I(u) &= \frac{1}{2}\|w\|^2 - \int_{\mathbb{A}_2} G_0(x,u)\,dx - \int_{\mathbb{A}_1} G_0(x,u)\,dx \\ &\ge C^{**}\|w\|^2 - 4(C_3)^2\epsilon\|w\|^2 - (2C_3)^{\lambda+1} C_\epsilon \|w\|^{s+1} - \int_{\mathbb{A}_1} G(x,u)\,dx \\ &\ge C_4\|w\|^2 - C_5\|w\|^{s+1}\end{aligned}$$

for all $u \in X^1$ with $\|u\| \le r_1$, which implies that

$$I(u) \ge 0, \quad \forall u \in X^1 \text{ with } \|u\| \le r$$

for 0<r small enough.

*(3)*We claim that *I* satisfies $(C)_c^*$. Consider a sequence (uαn) such that (uαn) is admissible and

$$u_{\alpha_n} \in X_{\alpha_n}, \quad I(u_{\alpha_n}) \to c, \qquad (1 + \|u_{\alpha_n}\|) I'(u_{\alpha_n}) \to 0 \tag{3.4}$$

and

$$\lim_{n\to\infty}\int\left(\frac{1}{2}g_0(x,u_{\alpha_n})u_{\alpha_n}-G_0(x,u_{\alpha_n})\right)=c. \tag{3.5}$$

Let $w_{\alpha_n}=\|u_{\alpha_n}\|^{-1}u_{\alpha_n}$. Up to a subsequence, we have

$$w_{\alpha_n}\rightharpoonup w \quad \text{in } X, \qquad w_{\alpha_n}\to w \quad \text{in } L^2_{\text{loc}}, \quad w_{\alpha_n}(x)\to w(x) \quad \text{a.e. } x\in\mathbb{R}^N.$$

If w=0, we choose a sequence {tn}⊂[0,1] such that

$$I(t_n u_{\alpha_n})=\max_{t\in[0,1]} I(tu_{\alpha_n}).$$

For any m $m>0$, let $v_{\alpha_n}=2\sqrt{m}w_{\alpha_n}$. Now, we claim that

$$\lim_{n\to\infty}\int G_0(x,v_{\alpha_n})=0.$$

Let ϵ>0; for r≥1, then,

$$\int_{|v_{\alpha_n}|\geq r} G_0(x,v_{\alpha_n})\,dx \leq C_6 r^{p-2^*}\int_{|v_{\alpha_n}|\geq r}|v_{\alpha_n}|^{2^*}\,dx$$
$$\leq C_7 r^{p-2^*}|v_{\alpha_n}|_{2^*}^{2^*}.$$

Since p<2∗, we may fix *r* large enough such that

$$\left|\int_{|v_{\alpha_n}|\geq r} G_0(x,v_{\alpha_n})\,dx\right|\leq\frac{\epsilon}{3}$$

for all *n*. Moreover, by (H_6), there exists R>0 $R>0$ such that

$$\left|\int_{|v_{\alpha_n}|\geq r} G_0(x,v_{\alpha_n})\,dx\right|\leq |v_{\alpha_n}|_2^2\sup_{|t|\leq r,|x|\geq R}\frac{|G_0(x,t)|}{t^2}\leq\frac{\epsilon}{3}$$

for all *n*. Finally, since vαn→0 in Ls(BR(0)) for s∈[2,2∗), we can use (H_1) again to derive

$$\left|\int_{|v_{\alpha_n}|\leq m|x|\leq R} G_0(x,v_{\alpha_n})\,dx\right|\leq\frac{\epsilon}{3}$$

for *n* large enough. Combining the above three formulas, our claim holds.

So, for *n* large enough, $2\sqrt{m}\|u_{\alpha_n}\|^{-1}\in(0,1)$, we have

$$I(t_nu_{\alpha_n})\geq I(v_{\alpha_n})\geq m-\epsilon\geq\frac{m}{2},\tag{3.6}$$

where ϵ is a small enough constant.

That is, $I(t_nu_{\alpha_n})\to\infty$. Now, $I(0)=0$, $I(u_{\alpha_n})\to c$, we know that tn∈[0,1] and

$$\int\left(\left|\nabla(t_nu_{\alpha_n})\right|^2-V(x)t_n^2|u_{\alpha_n}|^2\right)$$
$$+\lambda\int t_n^2|u_{\alpha_n}|^2-\int g_0(x,t_nu_{\alpha_n})t_nu_{\alpha_n}=t_n\frac{d}{dt}\bigg|_{t=t_n}I(tu_{\alpha_n})=0.\tag{3.7}$$

Therefore, using (H_4), we have

$$\int\frac{1}{2}g_0(x,u_{\alpha_n})u_{\alpha_n}-G_0(x,u_{\alpha_n})\geq\frac{1}{\theta}\int\left(\frac{1}{2}g_0(x,t_nu_{\alpha_n})t_nu_{\alpha_n}-G_0(x,t_nu_{\alpha_n})\right)\to+\infty.$$

This contradicts (3.5).

If w≠0, then the set ⊖={x∈RN:w(x)≠0} has a positive Lebesgue measure. For x∈⊖, we have |uαn(x)|→∞. Hence, by (H_3), we have

$$\frac{g_0(x,u_{\alpha_n}(x))u_{\alpha_n}(x)}{|u_{\alpha_n}(x)|^2}\left|w_{\alpha_n}(x)\right|^2\to\infty.\tag{3.8}$$

From (3.4), we obtain

$$1 - o(1) \geq \left(\int_{w \neq 0} + \int_{w=0} \right) \frac{g_0(x, u_{\alpha_n}(x)) u_{\alpha_n}(x)}{|u_{\alpha_n}(x)|^2} \left| w_{\alpha_n}(x) \right|^2 dx. \tag{3.9}$$

By (3.8), the right-hand side of (3.9) →+∞. This is a contradiction.

In any case, we obtain a contradiction. Therefore, {uαn} is bounded.

Next, we denote {uαn} as {un} and prove {un} contains a convergent subsequence.

In fact, we know that {un} is bounded in *X*. Passing to a subsequence, we may assume that un⇀u in *X*. In order to establish strong convergence, it suffices to show that

$$\|u_n\| \to \|u\|.$$

By the condition (H_6) and ⟨I′(un),un−u⟩→0, we can similarly conclude it according to the above proof of our claim.

Finally, we claim that for every m∈N,

$$I(u) \to -\infty \quad \text{as } \|u\| \to \infty, u \in X_m^1 \oplus X^2.$$

By (H_2) and (H_3), there exist large enough *M* and some positive constant *T* such that

$$G_0(x,t) \geq Mt^2, \quad x \in \mathbb{R}^N, t \geq T.$$

So, for any $u \in X_m^1 \oplus X^2$, we have

$$\begin{aligned} I(tu) &= \frac{1}{2}t^2 \int \left(|\nabla u|^2 - V(x)|u|^2 \right) + \frac{t^2}{2}\lambda \int u^2 - \int G_0(x, tu) \\ &\leq \frac{1}{2}t^2 \int \left(|\nabla u|^2 - V(x)|u|^2 \right) + \frac{t^2}{2}\lambda \int u^2 - Mt^2 \int u^2 \to -\infty \quad \text{as } t \to +\infty. \end{aligned}$$

Hence, our claim holds. □

Proof of Theorem 1.2 We omit the proof which depends on Lemma 2.2 and is similar to the preceding one since our result is a variant of Ding Yanheng's Theorem 1.2 (see [12]). □

ACKNOWLEDGEMENTS

The authors would like to thank the referees for valuable comments and suggestions in improving this paper. This work was supported by the National NSF (Grant No. 10671156) of China.

COMPETING INTERESTS

The authors declare that they have no competing interests.

AUTHORS' CONTRIBUTIONS

The authors read and approved the final manuscript.

REFERENCES

1. Bartolo P, Benci V, Fortunato D: Abstract critical point theorems and applications to some nonlinear problems with strong resonance at infinity. *Nonlinear Anal.* 1983, 7: 981–1012.

2. Capozzi A, Lupo D, Solimini S: On the existence of a nontrivial solution to nonlinear problems at resonance. *Nonlinear Anal.* 1989, 13: 151–163.

3. Costa DG, Silva EA: On a class of resonant problems at higher eigenvalues. *Differ. Integral Equ.* 1995, 8: 663–671.

4. Costa DG, Tehrani HT:On a class of asymptotically linear elliptic problems in RN$\mathbb{R}^N$. *J. Differ. Equ.* 2001, 173: 470–494.

5. Stuart CA, Zhou HS:Applying the mountain pass theorem to an asymptotically linear elliptic equation on RN$\mathbb{R}^N$. *Commun. Partial Differ. Equ.*1999, 24: 1731–1758.

6. Tehrani HT:A note on asymptotically linear elliptic problems in RN$\mathbb{R}^N$. *J. Math. Anal. Appl.* 2002, 271: 546–554.

7. Li SJ, Willem M: Applications of local linking to critical point theory. *J. Math. Anal. Appl.* 1995, 189: 6–32.

8. Wang ZP, Zhou HS:Positive solutions for a nonhomogeneous elliptic equation on RN$\mathbb{R}^N$ without (AR) condition. *J. Math. Anal. Appl.* 2009, 353: 470–479.

9. Liu CY, Wang ZP, Zhou HS: Asymptotically linear Schrödinger equation with potential vanishing at infinity. *J. Diffcr. Equ.* 2008, 245: 201–222.

10. Jeanjean L, Tanaka K:A positive solution for a nonlinear Schrödinger equation on RN$\mathbb{R}^N$. *Indiana Univ. Math. J.* 2005, 54: 443–464.

11. Teng KM: Existence and multiplicity results for some elliptic systems with discontinuous nonlinearities. *Nonlinear Anal.* 2012, 75: 2975–2987.

12. Ding YH:Some existence results of solutions for the semilincar elliptic equations on RN. *J. Differ. Equ.* 1995, 119: 401–425.

8

MULTIPLE POSITIVE SOLUTIONS OF SEMILINEAR ELLIPTIC EQUATIONS INVOLVING CONCAVE AND CONVEX NONLINEARITIES IN $\mathbb{R}^N$

Huei-li Lin

Department of Natural Sciences in the Center for General Education, Chang Gung University, Tao-Yuan 333, Taiwan

ABSTRACT

In this article, we investigate the effect of the coefficient $f(z)$ of the sub-critical nonlinearity. For sufficiently large $\lambda > 0$, there are at least $k + 1$ positive solutions of the semilinear elliptic equations

$$\begin{cases} -\Delta v + \lambda v = f(z)v^{p-1} + h(z)v^{q-1} \text{ in } \mathbb{R}^N; \\ v \in H^1(\mathbb{R}^N), \end{cases}$$

where $1 \leq q < 2 < p < 2^* = 2N/(N-2)$ for $N \geq 3$.

KEYWORDS

semilinear elliptic equations, concave and convex positive solutions

1 INTRODUCTION

For $N \geq 3$, $1 \leq q < 2 < p < 2^* = 2N/(N-2)$, we consider the semilinear elliptic equations

$$\begin{cases} -\Delta v + \lambda v = f(z)v^{p-1} + h(z)v^{q-1} \text{ in } \mathbb{R}^N; \\ v \in H^1(\mathbb{R}^N), \end{cases} \qquad (E_\lambda)$$

where $\lambda > 0$.

Let f and h satisfy the following conditions:

$(f1)$ f is a positive continuous function in $\mathbb{R}^N$ and $\lim_{|z|\to\infty} f(z) = f_\infty > 0$.

$(f2)$ there exist k points $a^1, a^2, \ldots, a^k$ in $\mathbb{R}^N$ such that

$$f(a^i) = f_{\max} = \max_{z\in\mathbb{R}^N} f(z) \text{ for } 1 \leq i \leq k,$$

and $f_\infty < f_{\max}$.

$(h1)$ $h \in L^{\frac{p}{p-q}}(\mathbb{R}^N) \cap L^\infty(\mathbb{R}^N)$ and $h \gneqq 0$.

Semilinear elliptic problems involving concave-convex nonlinearities in a bounded domain

$$\begin{cases} -\Delta u = ch(z)|u|^{q-2}u + |u|^{p-2}u \text{ in } \Omega; \\ u \in H_0^1(\Omega), \end{cases} \qquad (E_c)$$

have been studied by Ambrosetti et al. [1] ($h \equiv 1$, and $1 < q < 2 < p \leq 2^* = 2N/(N-2)$) and Wu [2] $h \in C(\bar{\Omega})$ and changes sign, $1 < q < 2 < p < 2^*$). They proved that this equation has at least two positive solutions for sufficiently small $c > 0$. More general results of Equation (E_c) were done by Ambrosetti et al. [3], Brown and Zhang [4], and de Figueiredo et al. [5].

In this article, we consider the existence and multiplicity of positive solutions of Equation (E_λ) in $\mathbb{R}^N$. For the case $q = \lambda = 1$ and $f(z) \equiv 1$ for all $z \in \mathbb{R}^N$, suppose that h is nonnegative, small, and exponential decay, Zhu [6] showed that Equation (E_λ) admits at least two positive solutions in $\mathbb{R}^N$. Without the condition of exponential decay, Cao and Zhou [7] and Hirano [8] proved that Equation (E_λ) admits at least two positive solutions in $\mathbb{R}^N$. For the case $q = \lambda = 1$, by using the idea of category and Bahri-Li's minimax argument, Adachi and Tanaka [9] asserted that Equation (E_λ) admits at least four positive solutions in $\mathbb{R}^N$, where $f(z) \not\equiv 1$, $f(z) \geq 1 - C\exp((-(2+\delta)|z|)$ for some $C, \delta > 0$, and sufficiently small $\|h\|_{H^{-1}} > 0$. Similarly, in Hsu and Lin [10], they have studied that there are at least four positive solutions of the general case $-\Delta u + u = f(z)v^{p-1} + \lambda h(z)v^{q-1}$ in $\mathbb{R}^N$ for sufficiently small $\lambda > 0$.

By the change of variables

$$\varepsilon = \lambda^{-\frac{1}{2}} \text{ and } u(z) = \varepsilon^{\frac{2}{p-2}} v(\varepsilon z),$$

Equation (E_λ) is transformed to

$$\begin{cases} -\Delta u + u = f(\varepsilon z)u^{p-1} + \varepsilon^{\frac{2(p-q)}{p-2}} h(\varepsilon z)u^{q-1} \text{ in } \mathbb{R}^N; \\ u \in H^1(\mathbb{R}^N), \end{cases} \qquad (E_\varepsilon)$$

Associated with Equation (E_ε), we consider the C^1-functional J_ε, for $u \in H^1(\mathbb{R}^N)$,

$$J_\varepsilon(u) = \frac{1}{2}\|u\|_H^2 - \frac{1}{p}\int_{\mathbb{R}^N} f(\varepsilon z)u_+^p\,dz - \frac{1}{q}\int_{\mathbb{R}^N} \varepsilon^{\frac{2(p-q)}{p-2}} h(\varepsilon z)u_+^q\,dz,$$

where $\|u\|_H^2 = \int_{\mathbb{R}^N} \left(|\Delta u|^2 + |u|^2\right) dz$ is the norm in $H^1(\mathbb{R}^N)$ and $u_+ = \max\{u, 0\} \geq 0$. We know that the nonnegative weak solutions of Equation (E_ε) are equivalent to the critical points of J_ε. This article is organized as follows. First of all, we use the argument of Tarantello [11] to divide the Nehari manifold $\mathbf{M}_\varepsilon$ into the two parts $\mathbf{M}_\varepsilon^+$ and $\mathbf{M}_\varepsilon^-$. Next, we prove that the existence of a positive ground state solution $u_0 \in \mathbf{M}_\varepsilon^+$ of Equation (E_ε). Finally, in Section 4, we show that the condition ($f2$) affects the number of positive solutions of Equation (E_ε), that is, there are at least k critical points $u_1, \ldots, u_k \in \mathbf{M}_\varepsilon^-$ of J_ε such that $J_\varepsilon(u_i) = \beta_\varepsilon^i((PS)-\text{value})$ for $1 \leq i \leq k$.

Let

$$S = \sup_{\substack{u\in H^1(\mathbb{R}^N)\\ \|u\|_H=1}} \|u\|_{L^p},$$

Then

$$\|u\|_{L^p} \le S\|u\|_H \quad \text{for any } u \in H^1(\mathbb{R}^N)\setminus\{0\}. \tag{1.1}$$

For the semilinear elliptic equations

$$\begin{cases} -\Delta u + u = f(\varepsilon z)u^{p-1} \text{ in } \mathbb{R}^N; \\ u \in H^1(\mathbb{R}^N), \end{cases} \qquad (E0)$$

we define the energy functional $I_\varepsilon(u) = \frac{1}{2}\|u\|_H^2 - \frac{1}{p}\int_{\mathbb{R}^N} f(\varepsilon z)u_+^p\,dz$, and

$$\gamma_\varepsilon = \inf_{u\in \mathbf{N}_\varepsilon} I_\varepsilon(u),$$

where $\mathbf{N}_\varepsilon = \{u \in H^1(\mathbb{R}^N)\setminus\{0\} \mid u_+ \not\equiv 0 \text{ and } \langle I'_\varepsilon(u), u\rangle = 0\}$. Note that

(*i*) if $f \equiv f_\infty$, we define $I_\infty(u) = \frac{1}{2}\|u\|_H^2 - \frac{1}{p}\int_{\mathbb{R}^N} f_\infty u_+^p\,dz$ and

$$\gamma_\infty = \inf_{u\in \mathbf{N}_\infty} I_\infty(u),$$

where $\mathbf{N}_\infty = \{u \in H^1(\mathbb{R}^N)\setminus\{0\} \mid u_+ \not\equiv 0 \text{ and } \langle I'_{\max}(u), u\rangle = 0\}$;

(*ii*) if $f \equiv f_{\max}$, we define $I_{\max}(u) = \frac{1}{2}\|u\|_H^2 - \frac{1}{p}\int_{\mathbb{R}^N} f_{\max}u_+^p\,dz$ and

$$\gamma_{\max} = \inf_{u\in \mathbf{N}_{\max}} I_{\max}(u),$$

where $\mathbf{N}_{\max} = \{u \in H^1(\mathbb{R}^N)\setminus\{0\} \mid u_+ \not\equiv 0 \text{ and } \langle I'_{\max}(u), u\rangle = 0\}$.

Lemma 1.1

$$\gamma_{\max} = \frac{p-2}{2p}\left(f_{\max}S^p\right)^{-2/(p-2)} > 0.$$

Proof. It is similar to Theorems 4.12 and 4.13 in Wang [[12], p. 31].

Our main results are as follows.

(*I*) Let $\Lambda = \varepsilon^{2(p-q)/(p-2)}$. Under assumptions ($f$ 1) and (h 1), if

$$0 < \Lambda < \Lambda_0 = (p-2)\left(\frac{2-q}{f_{max}}\right)^{\frac{2-q}{p-2}}\left[(p-q)S^2\right]^{\frac{q-p}{p-2}}\|h\|_{\#}^{-1},$$

where $\|h\|_{\#}$ is the norm in $L^{\frac{p}{p-q}}(\mathbb{R}^N)$then Equation ($E_\varepsilon$) admits at least a positive ground state solution. (See Theorem 3.4)

(*II*) Under assumptions (f 1) - (f 2) and (h 1), if λ is sufficiently large, then Equation (E_λ) admits at least k + 1 positive solutions. (See Theorem 4.8)

2 THE NEHARI MANIFOLD

First of all, we define the Palais-Smale (denoted by (PS)) sequences and (PS)-conditions in $H^1(\mathbb{R}^N)$ for some functional J.

Definition 2.1 (*i*) *For $\beta \in \mathbb{R}$, a sequence $\{u_n\}$ is a (PS)$_\beta$-sequence in $H^1(\mathbb{R}^N)$ for J if $J(u_n) = \beta + o_n(1)$ and $J'(u_n) = o_n(1)$strongly in $H^{-1}(\mathbb{R}^N)$ as $n \to \infty$, where $H^{-1}(\mathbb{R}^N)$ is the dual space of $H^1(\mathbb{R}^N)$;*

(*ii*) *J satisfies the (PS)$_\beta$-condition in $H^1(\mathbb{R}^N)$ if every (PS)$_\beta$-sequence in $H^1(\mathbb{R}^N)$ for J contains a convergent subsequence.*

Next, since J_ε is not bounded from below in $H^1(\mathbb{R}^N)$, we consider the Nehari manifold

$$\mathbf{M}_\varepsilon = \left\{u \in H^1(\mathbb{R}^N)\setminus\{0\} \mid u_+ \not\equiv 0 \text{ and } \langle J'_\varepsilon(u), u\rangle = 0\right\}, \tag{2.1}$$

where

$$\langle J'_\varepsilon(u), u\rangle = \|u\|_H^2 - \int_{\mathbb{R}^N} f(\varepsilon z)u_+^p\,dz - \int_{\mathbb{R}^N} \varepsilon^{\frac{2(p-q)}{p-2}} h(\varepsilon z)u_+^q\,dz.$$

Note that $\mathbf{M}_\varepsilon$ contains all nonnegative solutions of Equation (E_ε). From the lemma below, we have that J_ε is bounded from below on $\mathbf{M}_\varepsilon$.

Lemma 2.2 *The energy functional J_ε is coercive and bounded from below on $\mathbf{M}_\varepsilon$.*

Proof. For $u \in \mathbf{M}_\varepsilon$, by (2.1), the Hölder inequality $\left(p_1 = \frac{p}{p-q}, p_2 = \frac{p}{q}\right)$ and the Sobolev embedding theorem (1.1), we get

$$\begin{aligned} J_\varepsilon(u) &= \left(\frac{1}{2} - \frac{1}{p}\right)\|u\|_H^2 - \left(\frac{1}{q} - \frac{1}{p}\right)\int_{\mathbb{R}^N} \varepsilon^{\frac{2(p-q)}{p-2}} h(\varepsilon z)u_+^q\,dz \\ &\geq \frac{\|u\|_H^q}{p}\left[\frac{p-2}{2}\|u\|_H^{2-q} - \frac{p-q}{q}\varepsilon^{\frac{2(p-q)}{p-2}}\|h\|_\# S^q\right]. \end{aligned}$$

Hence, we have that J_ε is coercive and bounded from below on $\mathbf{M}_\varepsilon$.

Define

$$\psi_\varepsilon(u) = \langle J'_\varepsilon(u), u\rangle.$$

Then for $u \in \mathbf{M}_\varepsilon$, we get

$$\begin{aligned} \langle \psi'_\varepsilon(u), u\rangle &= 2\|u\|_H^2 - p\int_{\mathbb{R}^N} f(\varepsilon z)u_+^p\,dz - q\int_{\mathbb{R}^N} \varepsilon^{\frac{2(p-q)}{p-2}} h(\varepsilon z)u_+^q\,dz \\ &= (p-q)\int_{\mathbb{R}^N} \varepsilon^{\frac{2(p-q)}{p-2}} h(\varepsilon z)u_+^q\,dz - (p-2)\|u\|_H^2 \end{aligned} \tag{2.2}$$

$$= (2-q)\|u\|_H^2 - (p-q)\int_{\mathbb{R}^N} f(\varepsilon z)u_+^p\,dz. \tag{2.3}$$

We apply the method in Tarantello [11], let

$\mathbf{M}_\varepsilon^+ = \{u \in \mathbf{M}_\varepsilon \mid \langle \psi'_\varepsilon(u), u\rangle > 0\};$
$\mathbf{M}_\varepsilon^0 = \{u \in \mathbf{M}_\varepsilon \mid \langle \psi'_\varepsilon(u), u\rangle = 0\};$
$\mathbf{M}_\varepsilon^- = \{u \in \mathbf{M}_\varepsilon \mid \langle \psi'_\varepsilon(u), u\rangle < 0\}.$

Lemma 2.3 *Under assumptions* $(f1)$ *and* $(h1)$, *if* $0 < \Lambda\ (= \varepsilon^{2(p-q)/(p-2)}) < \Lambda_0$, *then* $\mathbf{M}_\varepsilon^0 = \emptyset$.

Proof. See Hsu and Lin [[10], Lemma 5].

Lemma 2.4 *Suppose that u is a local minimizer for* J_ε *on* $\mathbf{M}_\varepsilon$ *and* $u \notin \mathbf{M}_\varepsilon^0$. *Then* $J'_\varepsilon(u) = 0$ *in* $H^{-1}(\mathbb{R}^N)$.

Proof. See Brown and Zhang [[4], Theorem 2.3].

Lemma 2.5 *We have the following inequalities.*

(*i*) $\int_{\mathbb{R}^N} h(\varepsilon z)u_+^q\, dz > 0$ *for each* $u \in \mathbf{M}_\varepsilon^+$;

(*ii*) $\|u\|_H < \left(\frac{p-q}{p-2}\Lambda\|h\|_\# S^q\right)^{1/(2-q)}$ *for each* $u \in \mathbf{M}_\varepsilon^+$;

(*iii*) $\|u\|_H > \left[\frac{2-q}{(p-q)f_{max}S^p}\right]^{1/(p-2)}$ *for each* $u \in \mathbf{M}_\varepsilon^-$;

(*iv*) $0 < \Lambda \left(= \varepsilon^{2(p-q)/(p-2)}\right) < \frac{q\Lambda_0}{2}$, *then* $J_\varepsilon(u) > 0$ *for each* $u \in \mathbf{M}_\varepsilon^-$.

Proof. (*i*) It can be proved by using (2.2).

(*ii*) For any $u \in \mathbf{M}_\varepsilon^+ \subset \mathbf{M}_\varepsilon$, by (2.2), we apply the Hölder inequality $(p_1 = \frac{p}{p-q}, p_2 = \frac{p}{q})$ to obtain that

$$\begin{aligned} 0 &< (p-q)\int_{\mathbb{R}^N} \Lambda h(\varepsilon z)u_+^q\, dz - (p-2)\|u\|_H^2 \\ &\le (p-q)\Lambda\|h\|_\# S^q\|u\|_H^q - (p-2)\|u\|_H^2. \end{aligned}$$

(*iii*) For any $u \in \mathbf{M}_\varepsilon^-$, by (2.3), we have that

$$\|u\|_H^2 < \frac{p-q}{2-q}\int_{\mathbb{R}^N} f(\varepsilon z)u_+^p\, dz \le \frac{p-q}{2-q}S^p\|u\|_H^p f_{max}.$$

(*iv*) For any $u \in \mathbf{M}_\varepsilon^- \subset \mathbf{M}_\varepsilon$, by (*iii*), we get that

$$J_\varepsilon(u) = \left(\frac{1}{2}-\frac{1}{p}\right)\|u\|_H^2 - \left(\frac{1}{q}-\frac{1}{p}\right)\int_{\mathbb{R}^N} \Lambda h(\varepsilon z)u_+^q\,dz$$
$$\geq \frac{\|u\|_H^q}{p}\left[\frac{p-2}{2}\|u\|_H^{2-q} - \frac{p-q}{q}\Lambda\|h\|_\# S^q\right]$$
$$> \frac{1}{p}\left[\frac{2-q}{(p-q)f_{max}S^p}\right]^{\frac{q}{p-2}}\left[\frac{p-2}{2}\left[\frac{2-q}{(p-q)f_{max}S^p}\right]^{\frac{2-q}{p-2}} - \frac{p-q}{q}\Lambda\|h\|_\# S^q\right].$$

Thus, if $0 < \Lambda < \frac{q}{2}(p-2)\left(\frac{2-q}{f_{max}}\right)^{\frac{2-q}{p-2}}\left[(p-q)S^2\right]^{\frac{q-p}{p-2}}\|h\|_\#^{-1}$, we get that $J_\varepsilon(u) \geq d_0 > 0$ for some constant $d_0 = d_0(\varepsilon, p, q, S, \|h\|_\#, f_{max})$.

For $u \in H^1(\mathbb{R}^N) \setminus \{0\}$ and $u_+ \not\equiv 0$, let

$$\bar{t} = \bar{t}(u) = \left[\frac{(2-q)\|u\|_H^2}{(p-q)\int_{\mathbb{R}^N} f(\varepsilon z)u_+^p\,dz}\right]^{1/(p-2)} > 0.$$

Lemma 2.6 *For each $u \in H^1(\mathbb{R}^N) \setminus \{0\}$ and $u_+ \not\equiv 0$, we have that*

(*i*) *if* $\int_{\mathbb{R}^N} h(\varepsilon z)u_+^q\,dz = 0$, *then there exists a unique positive number* $t^- = t^-(u) > \bar{t}$ *such that* $t^-u \in \mathbf{M}_\varepsilon^-$ *and* $J_\varepsilon(t^-u) = \sup_{t\geq 0} J_\varepsilon(tu)$;

(*ii*) *if* $0 < \Lambda\ (= \varepsilon^{2(p-q)/(p-2)}) < \Lambda_0$ *and* $\int_{\mathbb{R}^N} h(\varepsilon z)u_+^q\,dz > 0$, *then there exist unique positive numbers* $t^+ = t^+(u) < \bar{t} < t^- = t^-(u)$ *such that* $t^+u \in \mathbf{M}_\varepsilon^+, t^-u \in \mathbf{M}_\varepsilon^-$ *and*

$$J_\varepsilon(t^+u) = \inf_{0\leq t\leq \bar{t}} J_\varepsilon(tu), \quad J_\varepsilon(t^-u) = \sup_{t\geq \bar{t}} J_\varepsilon(tu).$$

Proof. See Hsu and Lin [[10], Lemma 7].

Applying Lemma 2.3 ($\mathbf{M}_\varepsilon^0 = \emptyset$ for $0 < \Lambda < \Lambda_0$), we write $\mathbf{M}_\varepsilon = \mathbf{M}_\varepsilon^+ \cup \mathbf{M}_\varepsilon^-$, where

$$\mathbf{M}_\varepsilon^+ = \left\{u \in \mathbf{M}_\varepsilon | (2-q)\|u\|_H^2 - (p-q)\int_{\mathbb{R}^N} f(\varepsilon z)u_+^p\,dz > 0\right\},$$
$$\mathbf{M}_\varepsilon^- = \left\{u \in \mathbf{M}_\varepsilon | (2-q)\|u\|_H^2 - (p-q)\int_{\mathbb{R}^N} f(\varepsilon z)u_+^p\,dz < 0\right\}.$$

Define

$$\alpha_\varepsilon = \inf_{u \in \mathbf{M}_\varepsilon} J_\varepsilon(u); \quad \alpha_\varepsilon^+ = \inf_{u \in \mathbf{M}_\varepsilon^+} J_\varepsilon(u); \quad \alpha_\varepsilon^- = \inf_{u \in \mathbf{M}_\varepsilon^-} J_\varepsilon(u).$$

Lemma 2.7 (*i*) *If* $0 < \Lambda \ (= \varepsilon^{2(p-q)/(p-2)}) < \Lambda_0$, *then* $\alpha_\varepsilon \le \alpha_\varepsilon^+ < 0$;

(*ii*) *If* $0 < \Lambda < q\,\Lambda_0/2$, *then* $\alpha_\varepsilon^- \ge d_0 > 0$ *for some constant* $d_0 = d_0\ (\varepsilon, p, q, S, \|h\|_{\#}, f_{\max})$.

Proof. (*i*) Let $u \in \mathbf{M}_\varepsilon^+$, by (2.2), we get

$$(p-2)\,\|u\|_H^2 < (p-q)\int_{\mathbb{R}^N} \Lambda h(\varepsilon z) u_+^q\, dz.$$

Then

$$\begin{aligned} J_\varepsilon(u) &= \left(\frac{1}{2}-\frac{1}{p}\right)\|u\|_H^2 - \left(\frac{1}{q}-\frac{1}{p}\right)\int_{\mathbb{R}^N} \Lambda h(\varepsilon z) u_+^q\, dz \\ &< \left[\left(\frac{1}{2}-\frac{1}{p}\right) - \left(\frac{1}{q}-\frac{1}{p}\right)\frac{p-2}{p-q}\right]\|u\|_H^2 \\ &= -\frac{(2-q)(p-2)}{2pq}\|u\|_H^2 < 0. \end{aligned}$$

By the definitions of α_ε and α+ε, we deduce that $\alpha_\varepsilon \le \alpha_\varepsilon^+ < 0$.

(*ii*) See the proof of Lemma 2.5 (*iv*).

Applying Ekeland's variational principle and using the same argument in Cao and Zhou [7] or Tarantello [11], we have the following lemma.

Lemma 2.8 (*i*) *There exists a* $(PS)_{\alpha_\varepsilon}$ *-sequence* $\{u_n\}$ *in* $\mathbf{M}_\varepsilon$ *for* J_ε;

(*ii*) *There exists* $(PS)_{\alpha_\varepsilon^+}$*-sequence* $\{u_n\}$ *in* $\mathbf{M}_\varepsilon^+$ *for* J_ε;

(*iii*) *There exists a* $(PS)_{\alpha_\varepsilon^-}$*-sequence* $\{u_n\}$ *in* $\mathbf{M}_\varepsilon^-$ *for* J_ε.

3 EXISTENCE OF A GROUND STATE SOLUTION

In order to prove the existence of positive solutions, we claim that J_ε satisfies the $(PS)_\beta$-condition in $H^1(\mathbb{R}^N)$ for $\beta \in \left(-\infty, \gamma_\infty - C_0\Lambda^{\frac{2}{2-q}}\right)$, where $\Lambda = \varepsilon^{2(p-q)/(p-2)}$ and C_0 is defined in the following lemma.

Lemma 3.1 *Assume that h satisfies* $(h\,1)$ *and* $0 < \Lambda\ (= \varepsilon^{2(p-q)/(p-2)}) < \Lambda_0$. *If* $\{u_n\}$ *is a* $(PS)_\beta$*-sequence in* $H^1(\mathbb{R}^N)$ *for* J_ε *with* $u_n \rightharpoonup u$ *weakly in* $H^1(\mathbb{R}^N)$, *then* $J'_\varepsilon(u) = 0$ *in* $H^{-1}(\mathbb{R}^N)$ *and* $J_\varepsilon(u) \geq -C_0\Lambda^{\frac{2}{2-q}} \geq -C'_0$, *where*

$$C_0 = (2-q)[(p-q)\|h\|_\# S^q]^{\frac{2}{2-q}} \Big/ \left[2pq(p-2)^{\frac{q}{2-q}}\right],$$

and

$$C'_0 = \left[(p-2)(2-q)^{\frac{p}{p-2}}\right] \Big/ \left\{2pq[f_{max}(p-q)]^{\frac{2}{p-2}} S^{\frac{2p}{p-2}}\right\}.$$

Proof. Since $\{u_n\}$ is a $(PS)_\beta$-sequence in $H^1(\mathbb{R}^N)$ for J_ε with $u_n \rightharpoonup u$ weakly in $H^1(\mathbb{R}^N)$, it is easy to check that $J'_\varepsilon(u) = 0$ in $H^{-1}(\mathbb{R}^N)$ and $u \geq 0$. Then we have $\langle J'_\varepsilon(u), u\rangle = 0$, that is, $\int_{\mathbb{R}^N} f(\varepsilon z)u^p dz = \|u\|_H^2 - \int_{\mathbb{R}^N} \Lambda h(\varepsilon z)u^q dz$.
Hence, by the Young inequality $\left(p_1 = \frac{2}{q} \text{ and } p_2 = \frac{2}{2-q}\right)$

$$\begin{aligned} J_\varepsilon(u) &= \left(\frac{1}{2} - \frac{1}{p}\right)\|u\|_H^2 - \left(\frac{1}{q} - \frac{1}{p}\right)\int_{\mathbb{R}^N} \Lambda h(\varepsilon z)u^q dz \\ &\geq \frac{p-2}{2p}\|u\|_H^2 - \frac{p-q}{pq}\Lambda\|h\|_\# S^q\|u\|_H^q \\ &\geq \frac{p-2}{2p}\|u\|_H^2 - \frac{p-2}{pq}\left[\frac{q\|u\|_H^2}{2} + \left(\frac{p-q}{p-2}\Lambda\|h\|_\# S^q\right)^{\frac{2}{2-q}}\frac{2-q}{2}\right] \\ &\geq -\frac{(p-2)(2-q)^{\frac{p}{p-2}}}{2pq[f_{max}(p-q)]^{\frac{2}{p-2}} S^{\frac{2p}{p-2}}}. \end{aligned}$$

Lemma 3.2 *Assume that f and h satisfy $(f1)$ and $(h1)$. If $0 < \Lambda\ (= \varepsilon^{2(p-q)/(p-2)}) < \Lambda_0$, then J_ε satisfies the $(PS)_\beta$-condition in $H^1(\mathbb{R}^N)$ for* $\beta \in \left(-\infty, \gamma_\infty - C_0\Lambda^{\frac{2}{2-q}}\right)$.

Proof. Let $\{u_n\}$ be a $(PS)_\beta$-sequence in $H^1(\mathbb{R}^N)$ for J_ε such that $J_\varepsilon(u_n) = \beta + o_n(1)$ and $J'_\varepsilon(u_n) = o_n(1)$ in $H^{-1}(\mathbb{R}^N)$. Then

$$|\beta| + c_n + \frac{d_n\|u_n\|_H}{p} \geq J_\varepsilon(u_n) - \frac{1}{p}\langle J'_\varepsilon(u_n), (u_n)\rangle$$
$$= \left(\frac{1}{2} - \frac{1}{p}\right)\|u_n\|_H^2 - \left(\frac{1}{q} - \frac{1}{p}\right)\int_{\mathbb{R}^N} \varepsilon^{\frac{2(p-q)}{p-2}} h(\varepsilon z)(u_n)_+^q dz$$
$$\geq \frac{p-2}{2p}\|u_n\|_H^2 - \frac{p-q}{pq}\Lambda\|h\|_\# S^q\|u_n\|_H^q,$$

where $c_n = o_n(1)$, $d_n = o_n(1)$ as $n \to \infty$. It follows that $\{u_n\}$ is bounded in $H^1(\mathbb{R}^N)$. Hence, there exist a subsequence $\{u_n\}$ and a nonnegative $u \in H^1(\mathbb{R}^N)$ such that $J'_\varepsilon(u) = 0$ in $H^{-1}(\mathbb{R}^N)$, $u_n \rightharpoonup u$ weakly in $H^1(\mathbb{R}^N)$, $u_n \rightharpoonup u$ a.e. in $\mathbb{R}^N$, $u_n \rightharpoonup u$ strongly in $L^s_{loc}(\mathbb{R}^N)$ for any $1 \leq s < 2^*$. Using the Brézis-Lieb lemma to get (3.1) and (3.2) below.

$$\int_{\mathbb{R}^N} f(\varepsilon z)(u_n - u)_+^p dz = \int_{\mathbb{R}^N} f(\varepsilon z)(u_n)_+^p dz - \int_{\mathbb{R}^N} f(\varepsilon z)u^p dz + o_n(1); \tag{3.1}$$

$$\int_{\mathbb{R}^N} h(\varepsilon z)(u_n - u)_+^q dz = \int_{\mathbb{R}^N} h(\varepsilon z)(u_n)_+^q dz - \int_{\mathbb{R}^N} h(\varepsilon z)u^q dz + o_n(1). \tag{3.2}$$

Next, claim that

$$\int_{\mathbb{R}^N} h(\varepsilon z)|u_n - u|^q dz \to 0 \text{ as } n \to \infty. \tag{3.3}$$

For any $\sigma > 0$, there exists $r > 0$ such that $\int_{[B^N(0;r)]^c} h(\varepsilon z)^{\frac{p}{p-q}} dz < \sigma$. By the Hölder inequality and the Sobolev embedding theorem, we get

$$\left|\int_{\mathbb{R}^N} h(\varepsilon z)|u_n - u|^q dz\right| \leq \int_{B^N(0;r)} h(\varepsilon z)|u_n - u|^q dz + \int_{[B^N(0;r)]^c} h(\varepsilon z)|u_n - u|^q dz$$

$$\leq \|h\|_{\#}\left(\int_{B^N(0;r)} |u_n - u|^p dz\right)^{q/p} + S^q\left(\int_{[B^N(0;r)]^c} h(\varepsilon z)^{\frac{p}{p-q}} dz\right)^{\frac{p-q}{p}} \|u_n - u\|_H^q$$

$$\leq C'\sigma + o_n(1).$$

$(\because \{u_n\}$ is bounded in $H^1(\mathbb{R}^N)$ and $u_n \to u$ in $L^p_{loc}(\mathbb{R}^N))$

Applying $(f1)$ and $u_n \to u$ in $L^p_{loc}(\mathbb{R}^N)$, we get that

$$\int_{\mathbb{R}^N} f(\varepsilon z)(u_n - u)_+^p dz = \int_{\mathbb{R}^N} f_\infty (u_n - u)_+^p dz + o_n(1). \tag{3.4}$$

Let $p_n = u_n - u$. Suppose $p_n \nrightarrow 0$ strongly in $H^1(\mathbb{R}^N)$. By (3.1)-(3.4), we deduce that

$$\|p_n\|_H^2 = \|u_n\|_H^2 - \|u\|_H^2 + o_n(1)$$

$$= \int_{\mathbb{R}^N} f(\varepsilon z)(u_n)_+^p dz - \int_{\mathbb{R}^N} \varepsilon^{\frac{2(p-q)}{p-2}} h(\varepsilon z)(u_n)_+^q dz$$

$$- \int_{\mathbb{R}^N} f(\varepsilon z)u^p dz + \int_{\mathbb{R}^N} \varepsilon^{\frac{2(p-q)}{p-2}} h(\varepsilon z)u^q dz + o_n(1)$$

$$= \int_{\mathbb{R}^N} f(\varepsilon z)(u_n - u)_+^p dz + o_n(1) = \int_{\mathbb{R}^N} f_\infty (p_n)_+^p dz + o_n(1).$$

Then

$$I_\infty(p_n) = \frac{1}{2}\|p_n\|_H^2 - \frac{1}{p}\int_{\mathbb{R}^N} f_\infty (p_n)_+^p dz$$

$$= \left(\frac{1}{2} - \frac{1}{p}\right)\|p_n\|_H^2 + o_n(1) > 0.$$

By Theorem 4.3 in Wang [12], there exists a sequence $\{s_n\} \subset \mathbb{R}^+$ such that $s_n = 1 + o_n(1)$, $\{s_n p_n\} \subset \mathbf{N}_\infty$ and $I_\infty(s_n p_n) = I_\infty(p_n) + o_n(1)$. It follows that

$$\begin{aligned}\gamma_\infty &\le I_\infty(s_n p_n) = I_\infty(p_n) + o_n(1)\\ &= J_\varepsilon(u_n) - J_\varepsilon(u) + o_n(1)\\ &= \beta - J_\varepsilon(u) + o_n(1) < \gamma_\infty,\end{aligned}$$

which is a contradiction. Hence, $u_n \to u$ strongly in $H^1(\mathbb{R}^N)$.

Remark 3.3 *By Lemma 1.1, we obtain*

$$C_0' = \frac{2-q}{q}\left(\frac{2-q}{p-q}\right)^{\frac{2}{p-2}} \gamma_{\max} < \gamma_{\max} < \gamma_\infty,$$

and $\gamma_\infty - C_0 \Lambda^{\frac{2}{2-q}} > 0$ *for* $0 < \Lambda < \Lambda_0$.

By Lemma 2.8 (*i*), there is a $(PS)_{\alpha_\varepsilon}$-sequence $\{u_n\}$ in $\mathbf{M}_\varepsilon$ for J_ε. Then we prove that Equation (E_ε) admits a positive ground state solution u_0 in $\mathbb{R}^N$.

Theorem 3.4 *Under assumptions* $(f1)$, $(h1)$, *if* $0 < \Lambda$ $(= \varepsilon^{2(p-q)/(p-2)}) <$ Λ_0, *then there exists at least one positive ground state solution* u_0 *of Equation* (E_ε) *in* $\mathbb{R}^N$. *Moreover, we have that* $u_0 \in \mathbf{M}_\varepsilon^+$ *and*

$$J_\varepsilon(u_0) = \alpha_\varepsilon = \alpha_\varepsilon^+ \ge -C_0 \Lambda^{\frac{2}{2-q}}. \tag{3.5}$$

Proof. By Lemma 2.8 (*i*), there is a minimizing sequence $\{u_n\} \subset \mathbf{M}_\varepsilon$ for J_ε such that $J_\varepsilon(u_n) = \alpha_\varepsilon + o_n(1)$ and $J_\varepsilon'(u_n) = o_n(1)$ in $H^{-1}(\mathbb{R}^N)$. Since $\alpha_\varepsilon < 0 < \gamma^\infty - C_0 \Lambda^{\frac{2}{2-q}}$, by Lemma 3.2, there exist a subsequence $\{u_n\}$ and $u_0 \in H^1(\mathbb{R}^N)$ such that $u_n \to u_0$ strongly in $H^1(\mathbb{R}^N)$. It is easy to see that u0≩0$u_0 \gneqq 0$ is a solution of Equation (E_ε) in $\mathbb{R}^N$ and $J_\varepsilon(u_0) = \alpha_\varepsilon$. Next, we claim that $u_0 \in \mathbf{M}_\varepsilon^+$. On the contrary, assume that $u_0 \in \mathbf{M}_\varepsilon^- (\mathbf{M}_\varepsilon^0 = \emptyset \text{ for } 0 < \Lambda (= \varepsilon^{2(p-q)/(p-2)}) < \Lambda_0)$.

We get that

$$\int_{\mathbb{R}^N} \Lambda h(\varepsilon z)(u_0)_+^q dz > 0.$$

Otherwise,

$$0 = \int_{\mathbb{R}^N} \Lambda h(\varepsilon z)(u_0)_+^q dz = \int_{\mathbb{R}^N} \Lambda h(\varepsilon z)(u_n)_+^q dz + o_n(1)$$
$$= \|u_n\|_H^2 - \int_{\mathbb{R}^N} f(\varepsilon z)(u_n)_+^p dz + o_n(1).$$

It follows that

$$\alpha_\varepsilon + o_n(1) = J_\varepsilon(u_n) = \left(\frac{1}{2} - \frac{1}{p}\right)\|u_n\|_H^2 + o_n(1),$$

which contradicts to $\alpha_\varepsilon < 0$. By Lemma 2.6 (*ii*), there exist positive numbers1$t^+ < \bar{t} < t^- = 1$ such that $t^+u_0 \in \mathbf{M}_\varepsilon^+, t^-u_0 \in \mathbf{M}_\varepsilon^-$ and

$$J_\varepsilon(t^+u_0) < J_\varepsilon(t^-u_0) = J_\varepsilon(u_0) = \alpha_\varepsilon,$$

which is a contradiction. Hence, $u_0 \in \mathbf{M}_\varepsilon^+$ and

$$-C_0\Lambda^{\frac{2}{2-q}} \le J_\varepsilon(u_0) = \alpha_\varepsilon = \alpha_\varepsilon^+.$$

By Lemma 2.4 and the maximum principle, then u_0 is a positive solution of Equation (E_ε) in $\mathbb{R}^N$.

4 EXISTENCE OF K+ 1 SOLUTIONS

From now, we assume that f and h satisfy $(f1)$-$(f2)$ and $(h1)$. Let $w \in H^1(\mathbb{R}^N)$ be the unique, radially symmetric, and positive ground state solution of Equation $(E0)$ in $\mathbb{R}^N$ for $f = f_{\max}$. Recall the facts (or see Bahri and Li [13], Bahri and Lions [14], Gidas et al. [15], and Kwong [16]).

(*i*) $w \in L^\infty(\mathbb{R}^N) \cap C_{loc}^{2,\theta}(\mathbb{R}^N)$ for some $0 < \theta < 1$ and $\lim_{|z|\to\infty} w(z) = 0$;

(*ii*) for any $\varepsilon > 0$, there exist positive numbers C_1, C_2^ε, and C_3^ε such that for all $z \in \mathbb{R}^N$

$$C_2^\varepsilon \exp(-(1-\varepsilon)|z|) \leq w(z) \leq C_1 \exp(-|z|)$$

and

$$|\nabla w(z)| \leq C_3^\varepsilon \exp(-(1-\varepsilon)|z|.)$$

For $1 \leq i \leq k$, we define

$$w_\varepsilon^i(z) = w\left(z - \frac{a^i}{\varepsilon}\right), \text{ where } f(a^i) = f_{max}.$$

Clearly, $w_\varepsilon^i(z) \in H^1(\mathbb{R}^N)$. By Lemma 2.6 (*ii*), there is a unique number $(t_\varepsilon^i)^- > 0$ such that $(t_\varepsilon^i)^- w_\varepsilon^i \in \mathbf{M}_\varepsilon^- \subset \mathbf{M}_\varepsilon$, where $1 \leq i \leq k$.

We need to prove that

$$\lim_{\varepsilon \to 0+} J_\varepsilon\left(\left(t_\varepsilon^i\right)^- w_\varepsilon^i\right) \leq \gamma_{max} \text{ uniformly in } i.$$

Lemma 4.1 (*i*) *There exists a number* $t_0 > 0$ *such that for* $0 \leq t \leq t_0$ *and any* $\varepsilon > 0$, *we have that*

$$J_\varepsilon\left(tw_\varepsilon^i\right) < \gamma_{max} \textit{ uniformly in } i;$$

(*ii*) *There exist positive numbers* t_1 *and* ε_1 *such that for any* $t > t_1$ *and* $\varepsilon < \varepsilon_1$, *we have that*

$$J_\varepsilon\left(tw_\varepsilon^i\right) < 0 \textit{ uniformly in } i.$$

Proof. (*i*) Since J_ε is continuous in $H^1(\mathbb{R}^N)$, $\{w_\varepsilon^i\}$ is uniformly bounded in $H^1(\mathbb{R}^N)$ for any $\varepsilon > 0$, and $\gamma_{max} > 0$, there is $t_0 > 0$ such that for $0 \leq t \leq t_0$ and any $\varepsilon > 0$

$$J_\varepsilon\left(tw_\varepsilon^i\right) < \gamma_{max}.$$

(*ii*) There is an $r_0 > 0$ such that $f(z) \geq f_{max}/2$ for $z \in B^N(a^i; r_0)$ uniformly in *i*. Then there exists $\varepsilon_1 > 0$ such that for $\varepsilon < \varepsilon_1$

$$J_\varepsilon\left(tw_\varepsilon^i\right) = \frac{t^2}{2}\left\|w_\varepsilon^i\right\|_H^2 - \frac{t^p}{p}\int_{\mathbb{R}^N} f(\varepsilon z)\left(w_\varepsilon^i\right)^p dz - \frac{t^q}{q}\int_{\mathbb{R}^N} \Lambda h(\varepsilon z)\left(w_\varepsilon^i\right)^q dz$$
$$\leq \frac{t^2}{2}\int_{\mathbb{R}^N}\left[|\nabla w|^2 w^2\right]dz - \frac{t^p}{2p}\int_{B^N(0;1)} f_{max} w^p dz.$$

Thus, there is $t_1 > 0$ such that for any $t > t_1$ and $\varepsilon < \varepsilon_1$

$J_\varepsilon\left(tw_\varepsilon^i\right) < 0$ uniformly in *i*.

Lemma 4.2 *Under assumptions* (f 1), (f 2), *and* (h 1). *If* $0 < \Lambda\ (= \varepsilon^{2(p-q)/(p-2)}) < q\,\Lambda_0/2$, *then*

$$\lim_{\varepsilon\to 0+}\sup_{t\to 0} J_\varepsilon\left(tw_\varepsilon^i\right) \leq \gamma_{max} \text{ uniformly in } i.$$

Proof. By Lemma 4.1, we only need to show that

$$\lim_{\varepsilon\to 0+}\sup_{t_0\leq t\leq t_1} J_\varepsilon\left(tw_\varepsilon^i\right) \leq \gamma_{max} \text{ uniformly in } i.$$

We know that $\sup_{t\geq 0} I_{max}(tw) = \gamma_{max}$. For $t_0 \leq t \leq t_1$, we get

$$J_\varepsilon\left(tw_\varepsilon^i\right) = \frac{1}{2}\left\|tw_\varepsilon^i\right\|_H^2 - \frac{1}{p}\int_{\mathbb{R}^N} f(\varepsilon z)\left(tw_\varepsilon^i\right)^p dz - \frac{1}{q}\int_{\mathbb{R}^N} \Lambda h(\varepsilon z)\left(tw_\varepsilon^i\right)^q dz$$
$$= \frac{t^2}{2}\int_{\mathbb{R}^N}\left[\left|\nabla w\left(z - \frac{a^i}{\varepsilon}\right)\right|^2 + w\left(z - \frac{a^i}{\varepsilon}\right)\right]dz$$
$$- \frac{t^p}{p}\int_{\mathbb{R}^N} f(\varepsilon z) w\left(z - \frac{a^i}{\varepsilon}\right)^p dz - \frac{t^q}{q}\int_{\mathbb{R}^N} \Lambda h(\varepsilon z) w\left(z - \frac{a^i}{\varepsilon}\right)^q dz$$
$$= \left\{\frac{t^2}{p}\int_{\mathbb{R}^N}\left[|\nabla w|^2 + w^2\right]dz - \frac{t^p}{p}\int_{\mathbb{R}^N} f_{max} w^p dz\right\}$$
$$+ \frac{t^p}{p}\int_{\mathbb{R}^N}(f_{max} - f(\varepsilon z)) w\left(z - \frac{a^i}{\varepsilon}\right)^p dz - \frac{t^q}{q}\Lambda\int_{\mathbb{R}^N} h(\varepsilon z) w\left(z - \frac{a^i}{\varepsilon}\right)^q dz$$
$$\leq \gamma_{max} + \frac{t_1^p}{p}\int_{\mathbb{R}^N}(f_{max} - f(\varepsilon z)) w\left(z - \frac{a^i}{\varepsilon}\right)^p dz - \frac{t_0^q}{q}\Lambda\int_{\mathbb{R}^N} h(\varepsilon z) w\left(z - \frac{a^i}{\varepsilon}\right) dz.$$

Since

$$\int_{\mathbb{R}^N} (f_{\max} - f(\varepsilon z)) w \left(z - \frac{a^i}{\varepsilon}\right)^p dz$$
$$= \int_{\mathbb{R}^N} \left[f_{\max} - f(\varepsilon z + a^i)\right] w^p dz = o(1) \text{ as } \varepsilon \to 0^+ \text{ uniformly in } i,$$

and

$$\Lambda \int_{\mathbb{R}^N} h(\varepsilon z) w \left(z - \frac{a^i}{\varepsilon}\right)^q dz \le \varepsilon^{\frac{2(p-q)}{p-2}} \|h\|_{\#} S^q \|w\|_H^q = o(1) \text{ as } \varepsilon \to 0^+,$$

then $\lim_{\varepsilon \to 0^+} \sup_{t_0 \le t \le t_1} J_\varepsilon\left(tw_\varepsilon^i\right) \le \gamma_{\max}$, that is, $\lim_{\varepsilon \to 0^+} \sup_{t \ge 0} J_\varepsilon\left(tw_\varepsilon^i\right) \le \gamma_{\max}$ uniformly in *i*.

Applying the results of Lemmas 2.6, 2.7(*ii*), and 4.2, we can deduce that

$$0 < d_0 \le \alpha_\varepsilon^- \le \gamma_{\max} + o(1) \text{ as } \varepsilon \to 0^+.$$

Since $\gamma_{\max} < \gamma_\infty$, there exists $\varepsilon_0 > 0$ such that

$$\gamma_{\max} < \gamma_\infty - C_0 \Lambda^{\frac{2}{2-q}} \text{ for any } \varepsilon < \varepsilon_0. \tag{4.1}$$

Choosing $0 < \rho_0 < 1$ such that

$$\overline{B_{\rho 0}^N(a^i)} \cap \overline{B_{\rho 0}^N(a^j)} = \emptyset \text{ for } i \ne j \text{ and } 1 \le i, j \le k,$$

where $\overline{B_{\rho 0}^N(a^i)} = \left\{z \in \mathbb{R}^N \mid \left|z - a^i\right| \le \rho 0\right\}$ and $f(a^i) = f_{\max}$. Define $\mathbf{K} = \{a^i \mid 1 \le i \le k\}$ and $\mathbf{K}_{\rho_0/2} = \cup_{i=1}^k \overline{B_{\rho_0/2}^N(a^i)}$. Suppose $\cup_{i=1}^k \overline{B_{\rho_0}^N(a^i)} \subset B_{r_0}^N(0)$ for some $r_0 > 0$.

Let $Q_\varepsilon : H^1(\mathbb{R}^N) \setminus \{0\} \to \mathbb{R}^N$ be given by

$$Q_\varepsilon(u) = \frac{\int_{\mathbb{R}^N} \chi(\varepsilon z) |u|^p dz}{\int_{\mathbb{R}^N} |u|^p dz},$$

where $\chi : \mathbb{R}^N \to \mathbb{R}^N$, $\chi(z) = z$ for $|z| \le r_0$ and $\chi(z) = r_0 z/|z|$ for $|z| > r_0$.

Lemma 4.3 *There exists* $0 < \varepsilon^0 \leq \varepsilon_0$ *such that if* $\varepsilon < \varepsilon^0$, *then* $Q_\varepsilon\left(\left(t_\varepsilon^i\right)^- w_\varepsilon^i\right) \in \mathbf{K}_{\rho_0/2}$ *for each* $1 \leq i \leq k$.

Proof. Since

$$Q_\varepsilon\left(\left(t_\varepsilon^i\right)^- w_\varepsilon^i\right) = \frac{\int_{\mathbb{R}^N} \chi(\varepsilon z)\left|w\left(z - \frac{a^i}{\varepsilon}\right)\right|^p dz}{\int_{\mathbb{R}^N}\left|w\left(z - \frac{a^i}{\varepsilon}\right)\right|^p dz} = \frac{\int_{\mathbb{R}^N} \chi(\varepsilon z + a^i)|w(z)|^p dz}{\int_{\mathbb{R}^N}|w(z)|^p dz} \to a^i \text{ as } \varepsilon \to 0^+,$$

there exists $\varepsilon^0 > 0$ such that

$$Q_\varepsilon\left(\left(t_\varepsilon^i\right)^- w_\varepsilon^i\right) \in \mathbf{K}_{\rho_0/2} \text{ for any } \varepsilon < \varepsilon^0 \text{ and each } 1 \leq i \leq k.$$

Lemma 4.4 *There exists a number* $\bar{\delta} > 0 \bar{\delta} > 0$ *such that if* $u \in \mathbf{N}_\varepsilon$ *and* $I_\varepsilon(u) \leq \gamma_{\max} + \bar{\delta}$, *then* $Q_\varepsilon(u) \in \mathbf{K}_{\rho_0/2}$ *for any* $0 < \varepsilon < \varepsilon^0$.

Proof. On the contrary, there exist the sequences $\{\varepsilon_n\} \subset \mathbb{R}^+$ and $\{u_n\} \subset \mathbf{N}_{\varepsilon_n}$ such that $\varepsilon_n \to 0$, $I_{\varepsilon_n}(u_n) = \gamma_{\max}(> 0) + o_n(1)$ as $n \to \infty$ and $Q_{\varepsilon_n}(u_n) \notin \mathbf{K}_{\rho_0/2}$ for all $n \in \mathbb{N}$. It is easy to check that $\{u_n\}$ is bounded in $H^1(\mathbb{R}^N)$. Suppose $u_n \to 0$ strongly in $L^p(\mathbb{R}^N)$. Since

$$\|u_n\|_H^2 = \int_{\mathbb{R}^N} f(\varepsilon_n z)(u_n)_+^p dz \text{ for each } n \in \mathbb{N},$$

and

$$I_{\varepsilon_n}(u_n) = \frac{1}{2}\|u_n\|_H^2 - \frac{1}{p}\int_{\mathbb{R}^N} f(\varepsilon_n z)(u_n)_+^p dz = \gamma_{\max} + o_n(1),$$

then

$$\gamma_{\max} + o_n(1) = I_{\varepsilon_n}(u_n) = \left(\frac{1}{2} - \frac{1}{p}\right)\int_{\mathbb{R}^N} f(\varepsilon_n z)(u_n)_+^p dz = o_n(1),$$

which is a contradiction. Thus, $u_n \nrightarrow 0$ strongly in $L^p(\mathbb{R}^N)$. Applying the concentration-compactness principle (see Lions [17] or Wang [[12], Lemma 2.16]), then there exist a constant $d_0 > 0$ and a sequence $\{\tilde{z}_n\} \subset \mathbb{R}^N$ such that

$$\int_{B^N(\tilde{z}_n;1)} |u_n(z)|^2 dz \geq d_0 > 0. \tag{4.2}$$

Let $v_n(z) = u_n(z + \tilde{z}_n)$, there are a subsequence $\{v_n\}$ and $v \in H^1(\mathbb{R}^N)$ such that $v_n \rightharpoonup v$ weakly in $H^1(\mathbb{R}^N)$. Using the similar computation in Lemma 2.6, there is a sequence $\{s^n_{max}\} \subset \mathbb{R}^+$ such that $\tilde{v}_n = s^n_{max} v_n \in \mathbf{N}_{max}$ and

$$\begin{aligned} 0 < \gamma_{max} &\leq I_{max}(\tilde{v}_n) \leq I_{\varepsilon_n}(s^n_{max} u_n) \\ &\leq I_{\varepsilon_n}(u_n) = \gamma_{max} + o_n(1) \text{ as } n \to \infty. \end{aligned}$$

We deduce that a convergent subsequence $\{s^n_{max}\}$ satisfies $s^n_{max} \to s_0 > 0$. Then there are subsequences $\{\tilde{v}_n\}$ and $\tilde{v} \in H^1(\mathbb{R}^N)$ such that $\tilde{v}_n \rightharpoonup \tilde{v}(= s_0 v)$ weakly in $H^1(\mathbb{R}^N)$. By (4.2), then $\tilde{v} \neq 0$. Moreover, we can obtain that vn˜→ṽstrongly in $H^1(\mathbb{R}^N)$ and $I_{max}(\tilde{v}) = \gamma_{max}$. Now, we want to show that there exists a subsequence $\{z_n\} = \{\varepsilon_n \tilde{z}_n\}$ such that $z_n \to z_0 \in \mathbf{K}$.

(*i*) Claim that the sequence $\{z_n\}$ is bounded in $\mathbb{R}^N$. On the contrary, assume that $|z_n| \to \infty$, then

$$\begin{aligned} \gamma_{max} &= I_{max}(\tilde{v}) < I_\infty(\tilde{v}) \\ &\leq \liminf_{n\to\infty} \left[\frac{1}{2}\|\tilde{v}_n\|_H^2 - \frac{1}{p}\int_{\mathbb{R}^N} f(\varepsilon_n z + z_n)(\tilde{v}_n)_+^p dz \right] \\ &= \liminf_{n\to\infty} \left[\frac{(s^n_{max})^2}{2}\|u_n\|_H^2 - \frac{(s^n_{max})^p}{p}\int_{\mathbb{R}^N} f(\varepsilon_n z)(u_n)_+^p dz \right] \\ &= \liminf_{n\to\infty} I_{\varepsilon_n}(s^n_{max} u_n) \leq \liminf_{n\to\infty} I_{\varepsilon_n}(u_n) = \gamma_{max}, \end{aligned}$$

which is a contradiction.

(*ii*) Claim that $z_0 \in \mathbf{K}$. On the contrary, assume that $z_0 \notin \mathbf{K}$, that is, $f(z_0) < f_{max}$. Then using the above argument to obtain that

$$\gamma_{max} = I_{max}(\tilde{v}) < \frac{1}{2}\|\tilde{v}\|_H^2 - \frac{1}{p}\int_{\mathbb{R}^N} f(z_0)(\tilde{v})_+^p\,dz$$
$$\leq \liminf_{n\to\infty}\left[\frac{1}{2}\|\widetilde{v_n}\|_H^2 - \frac{1}{p}\int_{\mathbb{R}^N} f(\varepsilon_n z + z_n)(\widetilde{v_n})_+^p\,dz\right]$$
$$= \gamma_{max},$$

which is a contradiction. Since $v_n \to v \neq 0$ *in* $H^1(\mathbb{R}^N)$, we have that

$$Q_{\varepsilon_n}(u_n) = \frac{\int_{\mathbb{R}^N} \chi(\varepsilon_n z)|v_n(z-\widetilde{z_n})|^p dz}{\int_{\mathbb{R}^N} |v_n(z-\widetilde{z_n})|^p dz}$$
$$= \frac{\int_{\mathbb{R}^N} \chi(\varepsilon_n z + \varepsilon_n \widetilde{z_n})|v_n|^p dz}{\int_{\mathbb{R}^N} |v_n|^p dz} \to z_0 \subset \mathbf{K}_{\rho_0/2} \text{ as } n \to \infty,$$

which is a contradiction.

Hence, there exists a number $\bar{\delta} > 0 \bar{\delta} > 0$ such that if u $I_\varepsilon(u) \leq \gamma_{max} + \bar{\delta}$, then $Q_\varepsilon(u) \in \mathbf{K}_{\rho_0/2}$ for any $0 < \varepsilon < \varepsilon^0$.

From (4.1), choosing $0 < \delta_0 < \bar{\delta}$ such that

$$\gamma_{max} + \delta_0 < \gamma_\infty - C_0\Lambda^{\frac{2}{2-q}} \text{ for any } 0 < \varepsilon < \varepsilon^0. \tag{4.3}$$

For each $1 \leq i \leq k$, define

$$O_\varepsilon^i = \{u \in \mathbf{M}_\varepsilon^- \mid |Q_\varepsilon(u) - a^i| < \rho_0\},$$
$$\partial O_\varepsilon^i = \{u \in \mathbf{M}_\varepsilon^- \mid |Q_\varepsilon(u) - a^i| = \rho_0\},$$

$$\beta_\varepsilon^i = \inf_{u\in O_\varepsilon^i} J_\varepsilon(u) \quad \text{and} \quad \tilde{\beta}_\varepsilon^i = \inf_{u\in \partial O_\varepsilon^i} J_\varepsilon(u).$$

Lemma 4.5 *If* $u \in \mathbf{M}_\varepsilon^-$ *and* $J_\varepsilon(u) \leq \gamma_{max} + \delta_0/2$, *then there exists a number* $0 < \bar{\varepsilon} < \varepsilon^0$ *such that* $2Q_\varepsilon(u) \in \mathbf{K}_{\rho_0/2}$ *for any* $0 < \varepsilon < \bar{\varepsilon}$.

Proof. We use the similar computation in Lemma 2.6 to get that there is a unique positive number

$$s_\varepsilon^u = \left(\frac{\|u\|_H^2}{\int\limits_{\mathbb{R}^N} f(\varepsilon z) u_+^p \, dz} \right)^{1/(p-2)}$$

such that $s_\varepsilon^u u \in \mathbf{N}_\varepsilon$. We want to show that $c^{s_\varepsilon^u} < c$ for some constant $c > 0$ (independent of u). First, since $u \in \mathbf{M}_\varepsilon^- \subset \mathbf{M}_\varepsilon$,

$$0 < d_0 \le \alpha_\varepsilon^- \le J_\varepsilon(u) \le \gamma_{\max} + \delta_0/2,$$

and J_ε is coercive on $\mathbf{M}_\varepsilon$, then $0 < c_2 < \|u\|_H^2 < c_1$ for some constants c_1 and c_2 (independent of u). Next, we claim that $\|u\|_{L^p}^p > c_3 > 0$ for some constant $c_3 > 0$ (independent of u). On the contrary, there exists a sequence $\{u_n\} \subset \mathbf{M}_\varepsilon^-$ such that

$\|u_n\|_{L^p}^p = o_n(1)$ as $n \to \infty$. By (2.3),

$$\frac{2-q}{p-q} < \frac{\int\limits_{\mathbb{R}^N} f(\varepsilon z)(u_n)_+^p \, dz}{\|u_n\|_H^2} \le \frac{f_{\max} \|u_n\|_{L^p}^p}{c_2} = o_n(1),$$

which is a contradiction. Thus, $s_\varepsilon^u < c$ for some constant $c > 0$ (independent of u). Now, we get that

$$\begin{aligned}
\gamma_{\max} + \delta_0/2 \ge J_\varepsilon(u) &= \sup_{t \ge 0} J_\varepsilon(tu) \ge J_\varepsilon(s_\varepsilon^u u) \\
&= \frac{1}{2}\|s_\varepsilon^u u\|_H^2 - \frac{1}{p}\int\limits_{\mathbb{R}^N} f(\varepsilon z)(s_\varepsilon^u u)_+^p \, dz - \frac{1}{q}\int\limits_{\mathbb{R}^N} \Lambda h(\varepsilon z)(s_\varepsilon^u u)_+^q \, dz \\
&\ge I_\varepsilon(s_\varepsilon^u u) - \frac{1}{q}\int\limits_{\mathbb{R}^N} \Lambda h(\varepsilon z)(s_\varepsilon^u u)_+^q \, dz.
\end{aligned}$$

From the above inequality, we deduce that

$$\begin{aligned}
I_\varepsilon(s_\varepsilon^u u) &\le \gamma_{\max} + \delta_0/2 + \frac{1}{q}\int\limits_{\mathbb{R}^N} \Lambda h(\varepsilon z)(s_\varepsilon^u u)_+^q \, dz \\
&\le \gamma_{\max} + \delta_0/2 + \Lambda \|h\|_\# S^q \|s_\varepsilon^u u\|_H^q \\
&< \gamma_{\max} + \delta_0/2 + \Lambda c^q (c_1)^{q/2} \|h\|_\# S^q, \text{ where } \Lambda = \varepsilon^{2(p-q)/(p-2)}.
\end{aligned}$$

Hence, there exists $0 < \bar{\varepsilon} < \varepsilon^0$ such that for $0 < \varepsilon < \bar{\varepsilon}$

$I_\varepsilon(s_\varepsilon^u u) \leq \gamma_{max} + \delta_0$, where $s_\varepsilon^u u \in \mathbf{N}_\varepsilon$.

By Lemma 4.4, we obtain

$$Q_\varepsilon(s_\varepsilon^u u) = \frac{\int_{\mathbb{R}^N} \chi(\varepsilon z)|s_\varepsilon^u u(z)|^p dz}{\int_{\mathbb{R}^N} |s_\varepsilon^u u(z)|^p dz} \in \mathbf{K}_{\rho_0/2} \text{ for any } 0 < \varepsilon < \bar{\varepsilon},$$

or $Q_\varepsilon(u) \in \mathbf{K}_{\rho_0/2}$ for any $0<\varepsilon<\bar{\varepsilon}$.

Applying the above lemma, we get that

$$\tilde{\beta}_\varepsilon^i \geq \gamma_{max} + \delta_0/2 \text{ for any } 0 < \varepsilon < \bar{\varepsilon}. \quad (4.4)$$

By Lemmas 4.2, 4.3, and Equation (4.3), there exists 0<ε∗≤ε̄ $0 < \varepsilon^* \leq \bar{\varepsilon}$ such that

$$\beta_\varepsilon^i \leq J_\varepsilon\left((t_\varepsilon^i)^- w_\varepsilon^i\right) \leq \gamma_{max} + \delta_0/3 < \gamma_\infty - C_0 \Lambda^{\frac{2}{2-q}} \text{ for any } 0 < \varepsilon < \varepsilon^*. \quad (4.5)$$

Lemma 4.6 *Given $u \in O_\varepsilon^i$, then there exist an $\eta > 0$ and a differentiable functional $l : B(0;\eta) \subset H^1(\mathbb{R}^N) \to \mathbb{R}^+$ such that $l(0) = 1$, $l(v)(u - v) \in O_\varepsilon^i$ for any $v \in B(0;\eta)$ and*

$$\langle l'(v), \phi\rangle|_{(l,v)=(1,0)} = \frac{\langle \psi'_\varepsilon(u), \phi\rangle}{\langle \psi'_\varepsilon(u), u\rangle} \text{ for any } \phi \in C_c^\infty(\mathbb{R}^N), \quad (4.6)$$

where $\psi_\varepsilon(u) = \langle J'_\varepsilon(u), u\rangle$.

Proof. See Cao and Zhou [7].

Lemma 4.7 *For each $1 \leq i \leq k$, there is a $(PS)_{\beta_\varepsilon^i}$-sequence $\{u_n\} \subset O_\varepsilon^i$ in $H^1(\mathbb{R}^N)$ for J_ε.*

Proof. For each $1 \leq i \leq k$, by (4.4) and (4.5),

$$\beta_\varepsilon^i < \tilde{\beta}_\varepsilon^i \text{ for any } 0 < \varepsilon < \varepsilon^*. \quad (4.7)$$

Then

$\beta^j_\varepsilon = \inf_{u \in O^j_\varepsilon \cup \partial O^j_\varepsilon} J_\varepsilon(u)$ for any $0 < \varepsilon < \varepsilon^*$.
.

Let $\{u^j_n\} \subset O^j_\varepsilon \cup \partial O^j_\varepsilon$ be a minimizing sequence for βiε. Applying Ekeland's variational principle, there exists a subsequence $\{u^j_n\}$ such that $J_\varepsilon(u^j_n) = \beta^j_\varepsilon + 1/n$ and

$$J_\varepsilon(u^j_n) \le J_\varepsilon(w) + \left\|w - u^j_n\right\|_H / n \text{ for all } w \in O^j_\varepsilon \cup \partial O^j_\varepsilon. \tag{4.8}$$

Using (4.7), we may assume that $u^j_n \in O^j_\varepsilon$ for sufficiently large *n*. By Lemma 4.6, then there exist an $\eta^j_n > 0$ and a differentiable functional lin: $+l^j_n : B(0; \eta^j_n) \subset H^1(\mathbb{R}^N) \to \mathbb{R}^+$ such that $l^j_n(0) = 1$, and $l^j_n(v)\left(u^j_n - v\right) \in O^j_\varepsilon$ for $v \in B(0; \eta^j_n)$. Let $v_\sigma = \sigma v$ with $\|v\|_H = 1$ and $0 < \sigma < \eta^j_n$. Then $v_\sigma \in B\left(0, \eta^j_n\right)$ and $w_\sigma = l^j_n(v_\sigma)\left(u^j_n - v_\sigma\right) \in O^j_\varepsilon$. From (4.8) and by the mean value theorem, we get that as $\sigma \to 0$

$$\begin{aligned}
\frac{\left\|w_\sigma - u^j_n\right\|_H}{n} &\ge J_\varepsilon(u^j_n) - J_\varepsilon(w_\sigma)\\
&= \langle J'_\varepsilon(t_0 u^j_n + (1 - t_0) w_\sigma), u^j_n - w_\sigma\rangle \text{ where } t_0 \in (0, 1)\\
&= \langle J'_\varepsilon(u^j_n), u^j_n - w_\sigma\rangle + o(\left\|u^j_n - w_\sigma\right\|_H) \left(\because J_\varepsilon \in C^1\right)\\
&= \sigma l^j_n(v_\sigma)\langle J'_\varepsilon(u^j_n), v\rangle + (1 - l^j_n(v_\sigma))\langle J'_\varepsilon(u^j_n), u^j_n\rangle + o(\left\|u^j_n - w_\sigma\right\|_H)\\
&\qquad (\because l^j_n(v_\sigma) \to l^j_n(0) = 1 \text{ as } \sigma \to 0)\\
&= \sigma l^j_n(\sigma v)\langle J'_\varepsilon(u^j_n), v\rangle + o(\left\|u^j_n - w_\sigma\right\|_H).
\end{aligned}$$

Hence,

$$\begin{aligned}
\left|\left\langle J'_\varepsilon\left(u^j_n\right), v\right\rangle\right| &\le \frac{\left\|w_\sigma - u^j_n\right\|_H \left(\frac{1}{n} + |o(1)|\right)}{\sigma\, |l^j_n(\sigma v)|}\\
&\le \frac{\left\|u^j_n\left(l^j_n(\sigma v) - l^j_n(0)\right) - \sigma v l^j_n(\sigma v)\right\|_H \left(\frac{1}{n} + |o(1)|\right)}{\sigma\, |l^j_n(\sigma v)|}\\
&\le \frac{\left\|u^j_n\right\|_H \left|l^j_n(\sigma v) - l^j_n(0)\right| + \sigma\|v\|_H \left|l^j_n(\sigma v)\right|}{\sigma\, |l^j_n(\sigma v)|} \left(\frac{1}{n} + |o(1)|\right)\\
&\le C\left(1 + \left\|\left(l^j_n\right)'(0)\right\|\right)\left(\frac{1}{n} + |o(1)|\right).
\end{aligned}$$

Since we can deduce that $\left\|\left(f_n^i\right)'(0)\right\| \leq c$ for all n and i from (4.6), then $J_\varepsilon'\left(u_n^i\right) = o_n(1)$ strongly in $H^{-1}(\mathbb{R}^N)$ as $n \to \infty$.

Theorem 4.8 *Under assumptions* $(f1)$, (f_2), *and* $(h1)$, *there exists a positive number* $\lambda^*(\lambda^* = (\varepsilon^*)^{-2})$ *such that for* $\lambda > \lambda^*$, *Equation*(E_λ) *has* $k + 1$ *positive solutions in* $\mathbb{R}^N$.

Proof. We know that there is a $(PS)_{\beta_\varepsilon^i}$-sequence $\{u_n\} \subset \mathbf{M}_\varepsilon^-$ in $H^1(\mathbb{R}^N)$ for J_ε for each $1 \leq i \leq k$, and (4.5). Since J_ε satisfies the $(PS)_\beta$-condition for $\beta \in \left(-\infty, \gamma_\infty - C_0\Lambda^{\frac{2}{2-q}}\right)$, then J_ε has at least k critical points in $\mathbf{M}_\varepsilon^-$ for $0 < \varepsilon < \varepsilon^*$. It follows that Equation (E_λ) has k nonnegative solutions in $\mathbb{R}^N$. Applying the maximum principle and Theorem 3.4, Equation (E_λ) has $k + 1$ positive solutions in $\mathbb{R}^N$.

ACKNOWLEDGEMENTS

The author was grateful for the referee's helpful suggestions and comments.

COMPETING INTERESTS

The author declares that he has no competing interests.

REFERENCES

1. Ambrosetti A, Brezis H, Cerami G: Combined effects of concave and convex nonlinearities in some elliptic problems. *J Funct Anal* 1994, 122: 519-543.

2. Wu TF: On semilinear elliptic equations involving concave-convex non-linearities and sign-changing weight function. *J Math Anal Appl*2006, 318: 253-270.

3. Ambrosetti A, Garcia Azorero J, Peral Alonso I: Multiplicity results for some nonlinear elliptic equations. *J Funct Anal* 1996, 137: 219-242.

4. Brown KJ, Zhang Y: The Nehari manifold for a semilinear elliptic equation with a sign-changing weight function. *J Diff Equ* 2003, 193: 481-499.

5. de Figueiredo DG, Gossez JP, Ubilla P: Local superlinearity and sub-linearity for indefinite semilinear elliptic problems. *J Funct Anal* 2003, 199: 452-467.

6. Zhu XP: A perturbation result on positive entire solutions of a semilinear elliptic equation. *J Diff Equ* 1991, 92: 163-178.

7. Cao DM, Zhou HS: Multiple positive solutions of nonhomogeneous semi-linear elliptic equations in $\mathbb{R}^N$. *Proc Roy Soc Edinburgh, Sect A*1996, 126: 443-463.

8. Hirano N: Existence of entire positive solutions for nonhomogeneous elliptic equations. *Nonlinear Anal* 1997, 29: 889-901.

9. Adachi S, Tanaka K: Four positive solutions for the semilinear elliptic equation: $-\Delta u + u = a(x)u^p + f(x)$ in $\mathbb{R}^N$. *Calc Var Partial Diff Equ*2000, 11: 63-95.

10. Hsu TS, Lin HL: Four positive solutions of semilinear elliptic equations involving concave and convex nonlinearities in $\mathbb{R}^N$. *J Math Anal Appl* 2010, 365: 758-775.

11. Tarantello G: On nonhomogeneous elliptic involving critical Sobolev exponent. *Ann Inst H Poincaré Anal Non Linéaire* 1992, 9: 281-304.

12. Wang HC: Palais-Smale approaches to semilinear elliptic equations in unbounded domains. *Electron J Diff Equ Monogragh* 2004, 142. 06

13. Bahri A, Li YY: On a min-max procedure for the existence of a positive solution for certain scalar field equations in $\mathbb{R}^N$. *Rev Mat Iberoamericana* 1990, 6: 1-15.

14. Bahri A, Lions PL: On the existence of a positive solution of semilin-ear elliptic equations in unbounded domains. *Ann Inst H Poincaré Anal Nonlinéaire* 1997, 14: 365-413.

15. Gidas B, Ni WM, Nirenberg L: Symmetry and related properties via the maximum principle. *Comm Math Phys* 1979, 68: 209-243.

16. Kwong MK: Uniqueness of positive solutions of $\Delta u - u + u^p = 0$ in $\mathbb{R}^N$. *Arch Ration Mech Anal* 1989, 105: 234-266.

17. Lions PL: The concentration-compactness principle in the calculus of variations. The locally compact case. *I II Ann Inst H Poincaré Anal Non-linéaire* 1984, 1: 109-145. 223-283

9

MULTIPLICITY OF SOLUTIONS FOR SINGULAR SEMILINEAR ELLIPTIC EQUATIONS IN WEIGHTED SOBOLEV SPACES

Gao Jia* and Long-jie Zhang

College of Science, University of Shanghai for Science and Technology, Shanghai, 200093, China

ABSTRACT

A class of semilinear elliptic equations involving strong resonance or non-resonance is reconsidered here. The multiplicity of solutions is investigated by using the variational method, and the results complement earlier ones.

KEYWORDS

multiplicity variational method, semilinear elliptic equations, weighted Sobolev space

1 INTRODUCTION

In this paper we investigate the following singular semilinear problem:

$$\mathcal{L}u - \lambda u\rho = -\rho f(x,u), \quad x \in \Omega, \tag{1.1}$$

where

$$\mathcal{L}u = -D_i\left(p_i^{1/2}p_j^{1/2}a_{ij}D_ju\right) + bu\rho, \tag{1.2}$$

and $\Omega \subseteq \mathbb{R}^N$ is an open set (possibly unbounded).

In recent years, the existence and multiplicity of solutions for semilinear equations have been extensively studied by many authors (see [1]–[11]).

In 2001, Shapiro [12] studied a series of eigenvalue problems of singular quasilinear elliptic and parabolic equations in weighted Sobolev spaces, he obtained many existence results by using Galerkin method. Then Jia *et al.* continued to study the quasilinear elliptic equations (see [13]–[15]). However, their main results are only concerned with the existence of solutions without considering multiplicity of solutions.

Motivated by previous work, the aim of this paper is to establish the existence of infinitely many solutions of the elliptic equations in weighted Sobolev spaces, by using the Ekeland variation principle and the mountain-pass lemma.

We first assume that

$(L-1)$ $a_{ij}(x), b(x) \in C^0(\Omega) \cap L^\infty(\Omega)$, $p_i(x), \rho(x) \in C^0(\Omega)$, $i,j = 1,2,\ldots,N$.
$(L-2)$ $p_i(x) > 0$, $\rho(x) > 0$, $\forall x \in \Omega$, and there exists $\varepsilon_0 > 0$ such that $b(x) \geq \varepsilon_0$, $\forall x \in \Omega$.
$(L-3)$ $\int_\Omega p_i < \infty$, $\int_\Omega \rho < \infty$.
$(L-4)$ $a_{ij}(x) = a_{ji}(x)$, $i,j = 1,2,\ldots,N$.
$(L-5)$ $a_{ij}\xi_i\xi_j \geq a_0|\xi|^2$, a_0 is a positive constant.

Let $\Gamma \subset \partial\Omega$ designate a fixed closed set (Γ may be empty).

We introduce the pre-Hilbert space:

$$C^1_{p,\rho}(\Omega,\Gamma)=\left\{u\in C^0(\overline{\Omega})\cap C^2(\Omega)\;\middle|\; u(x)=0,\forall x\in\Gamma;\int_\Omega\left(|D_iu|^2p_i+u^2\rho\right)<\infty\right\},$$

with the inner product

$$\langle u,v\rangle_{p,\rho}=\int_\Omega(D_iuD_ivp_i+uv\rho).$$

Let $H^1_{p,\rho}(\Omega,\Gamma)$ denote the Hilbert space completed by the norm $\|u\|_{p,\rho}=\langle u,u\rangle^{1/2}_{p,\rho}$, $L^2_\rho(\Omega)$ denote the Hilbert space with the inner product $\langle u,v\rangle_\rho=\int_\Omega uv\rho.$.

Next, we define the two-form

$$L(u,v)=\int_\Omega\left(p_i^{1/2}p_j^{1/2}a_{ij}D_iuD_jv+b\rho uv\right). \tag{1.3}$$

Definition 1.1

We say (Ω,Γ) is a V_L-region if the following two facts obtain:

(V_L-1) There exists a complete orthonormal system $\{\varphi_n\}_{n=1}^\infty$ in $L^2_\rho(\Omega)$, and $m(\{x\in\Omega\mid\varphi_i=0\})=0$. Also $\varphi_n\in H^1_{p,\rho}(\Omega,\Gamma)\cap C^2(\Omega)$.

(V_L-2) There exists a sequence of eigenvalues $\{\lambda_n\}_{n=1}^\infty$ with $0<\lambda_1\le\lambda_2\le\lambda_3\le\cdots\le\lambda_n\to\infty$ such that $L(\varphi_n,v)=\lambda_n\langle\varphi_n,v\rangle_\rho$, $\forall v\in H^1_{p,\rho}(\Omega,\Gamma)$. Also $\varphi_1>0$ in Ω.

Remark 1.1

$$\lambda_{k+1}=\inf\left\{L(u,u)\mid u\in V_k^\perp,\|u\|_{L^2_\rho}=1\right\},$$

$$L(u,u)\le\lambda_k\int_\Omega u^2\rho,\quad u\in V_k,k\ge1,$$

$$L(u,u)\ge\lambda_{k+1}\int_\Omega u^2\rho,\quad u\in V_k^\perp,k\ge0,$$

where Vk=span{φ1,φ2,...,φk}.

Definition 1.2

We say (Ω,Γ) is a simple VL-region if (Ω,Γ) is a VL-region and the following four conditions are satisfied:

(SV_L-1) $\Omega = \Omega_1 \times \Omega_2 \times \cdots \times \Omega_N$.

(SV_L-2) Associated with each Ω_i there are positive functions p_i^* and ρ_i^* in $C^0(\Omega)$ satisfying $\int_{\Omega_i}(p_i^* + \rho_i^*) < \infty$, for $i = 1, 2, \ldots, N$.

(SV_L-3) $\rho(x) = \rho_1^*(x_1)\cdots\rho_N^*(x_N)$ and

$$p_i(x) = \rho_1^*(x_1)\cdots\rho_{i-1}^*(x_{i-1})p_i^*(x_i)\rho_{i+1}^*(x_{i+1})\cdots\rho_N^*(x_N)$$

for i=1,...,N.

(SV_L-4) For each Ω_i $(i = 1, \ldots, N)$, $\exists h_i \in C^0(\Omega_i) \cap L^\theta_{\rho_i^*}(\Omega_i)$, for $2 < \theta < \infty$, with the property that

$$\forall\psi \in C^1(\Omega_i), \quad |\psi(s)| \le h_i(s)\|\psi\|_{p_i^*,\rho_i^*}, \quad \forall s \in \Omega_i,$$

where hi is in Lθρ∗i(Ωi) $L^\theta_{\rho_i^*}(\Omega_i)$ for $2 < \theta$, and also, to be quite explicit,

$$\|\psi\|^2_{p_i^*,\rho_i^*} = \int_{\Omega_i}\left(p_i^*(s)|\psi'(s)|^2 + \rho_i^*\psi^2(s)\right).$$

In fact, there are many examples which use special functions to illustrate the VL-region or the simple VL-region (see [12]).

The weak solutions of (1.1) correspond to the critical points of the functional Iλ(u): $H^1_{p,\rho}(\omega,\Gamma) \to \mathbb{R}$ defined by

$$I_\lambda(u) = \frac{1}{2}L(u,u) - \frac{\lambda}{2}\int_\Omega u^2\rho + \int_\Omega F(x,u)\rho, \quad u \in H^1_{p,\rho}(\Omega,\Gamma), \tag{1.4}$$

where $F(x,s) \in C^1(\overline{\Omega} \times \mathbb{R}, \mathbb{R})$ and $F(x,s) = \int_0^s f(x,t)\,dt$.

We define

$$\begin{cases} T^+ = \liminf_{s\to+\infty} F(x,s), & S^+ = \limsup_{s\to+\infty} F(x,s), \\ T^- = \liminf_{s\to-\infty} F(x,s), & S^- = \limsup_{s\to-\infty} F(x,s). \end{cases} \tag{1.5}$$

Here, the above functions belong to $L^1_\rho(\Omega)$ and the limits are taken a.e. and uniformly in x∈Ω.

As for f(x,s) and F(x,s), we make the assumptions as follows.

(F−1): There exists a function $k(x) > 0,\ k(x) \in L^1_\rho(\Omega),$ such that

$$\lim_{|s|\to+\infty} f(x,s) = 0, \qquad |F(x,s)| \le k(x), \quad \text{a.e. } x \in \Omega, \forall s \in \mathbb{R}.$$

(F−2): There exists a constant b1≥0 such that

$$F(x,s) \ge \frac{1}{2}(\lambda_1 - \lambda_2)s^2 + b_1\left(\int_\Omega \rho\right)^{-1}.$$

(F−3): $\int_\Omega S^+\rho \le 0$ and $\int_\Omega S^-\rho \le 0$.

(F−4): There exists t0∈R such that for the first eigenfunction φ1

$$\int_\Omega F(x, t_0\varphi_1)\rho < \min\left\{\int_\Omega T^+\rho, \int_\Omega T^-\rho\right\}.$$

(F−5): There exists $t_1^\pm \in \mathbb{R}$ such that for the first eigenfunction φ1

$$\int_\Omega F(x, t_1^\pm\varphi_1)\rho < \min\left\{\int_\Omega T^+\rho, \int_\Omega T^-\rho\right\}.$$

(F−6): There exist α∈(0,1) and δ>0 such that

$$F(x,s) \geq \frac{1-\alpha}{2}\lambda_1 s^2, \quad \forall x \in \Omega \text{ and } |s| < \delta.$$

(F′−1): Let η>2, f(x,−s)=−f(x,s) a.e. x∈RN, ∀s∈R and

$$0 > \frac{1}{\eta}F(x,s) \geq f(x,s)s, \quad \text{a.e. } x \in \mathbb{R}^N, \forall s \in \mathbb{R}.$$

(F′−2): There exists $1 < \theta < \frac{N+1}{N-1}$ such that

$$|f(x,s)| \leq C|s|^\theta, \quad \text{a.e. } x \in \mathbb{R}^N, \forall s \in \mathbb{R}$$

for some positive constant *C*.

$\int_{\Omega\cap\{\omega>0\}} S^+\rho + \int_{\Omega\cap\{\omega<0\}} S^-\rho \leq 0$, for every $\omega \in V(\lambda_k)$, where $V(\lambda_k)$ denotes the eigenspace associated with the eigenvalue λk.

(Fk−2): There exist $\gamma \in (\frac{\lambda_k}{2} - \frac{\lambda_1}{2}, \frac{\lambda_k}{2})$ and δ>0 $\delta > 0$ such that

$$F(x,s) \geq \gamma s^2, \quad \forall x \in \Omega \text{ and } |s| < \delta.$$

Remark 1.2

We must find the function F(x,s) satisfying (F−1)-(F−6). Let Ω=(0,π), λ1, and λ2 be the first eigenvalue and second eigenvalue of Laplacian eigenvalue problem of homogeneous boundary condition on Ω, and the first eigenfunction be ϕ1=sinx. Let $F(x,s) = \frac{1}{4}(\lambda_2 - \lambda_1)(2e^{-s^2} - 1)$. For $b_1 \equiv 0, \rho \equiv 1,$ For b1≡0, ρ≡1, by simple calculation, if t0 is large enough, we find that F(x,s) satisfies (F−1)-(F−6).

On the other hand, if we take $f(x,s) = -|s|^{k-1}s$, where $1 < k < \frac{N+1}{N-1}$, then f(x,s) satisfies (F′−1) and (F′−2).

The main results of our paper are given by the following theorems.

Theorem 1.1

Let$\Omega \subseteq R^N$ *be a domain satisfying the requirements of being a*VL-*region*. *Let*$\lambda=\lambda_1$*in* (1.1). *Assume that* $F(x,s) \in C^1(\overline{\Omega} \times \mathbb{R}, \mathbb{R})$. *Furthermore, if the conditions* (F−1), (F−3), *and* (F−4) *are satisfied, then the problem* (1.1)*has at least one nontrivial solution in* $H^1_{p,\rho}(\Omega, \Gamma)$.

Theorem 1.2

Let$\Omega \subseteq R^N$ *be a domain satisfying*VL-*region*. *And*$\lambda=\lambda_1$*in* (1.1). *Assume that* $F(x,s) \in C^1(\overline{\Omega} \times \mathbb{R}, \mathbb{R})$. *Furthermore, if the conditions* (F−1)-(F−3) *and* (F−5) *are satisfied, then the problem* (1.1) *has at least two nontrivial solutions in* $H^1_{p,\rho}(\Omega, \Gamma)$.

Theorem 1.3

Let$\Omega \subseteq$RN*be a domain satisfying simple*VL-*region*, (aij)=E, *and*$\lambda=\lambda_1$. *Assume that* $F(x,s) \in C^1(\overline{\Omega} \times \mathbb{R}, \mathbb{R})$.*Furthermore, if the conditions* (F−1), (F−3), (F−4), *and* (F−6) *are satisfied, then the problem* (1.1) *has at least two nontrivial solutions in* $H^1_{p,\rho}(\Omega, \Gamma)$.

Theorem 1.4

Let $\Omega \subseteq R^N$ *be a domain satisfying simple*VL-*region*, (aij)=E, *and*$\lambda=\lambda_1$. *Assume that* $F(x,s) \in C^1(\overline{\Omega} \times \mathbb{R}, \mathbb{R})$.*Furthermore, if the conditions* (F−1)-(F−3), (F−5), *and* (F−6) *are satisfied, then the problem* (1.1) *has at least three nontrivial solutions in* $H^1_{p,\rho}(\Omega, \Gamma)$.

Theorem 1.5

Let$\Omega \subseteq$RN*be a domain satisfying simple*VL-*region*, $\lambda<\lambda_1$*and*(aij)=E. *Assume that* $F(x,s) \in C^1(\overline{\Omega} \times \mathbb{R}, \mathbb{R})$.*Furthermore, suppose the conditions* (F′−1) *and* (F′−2). *Then the problem* (1.1) *has infinitely many solutions in* $H^1_{p,\rho}(\Omega, \Gamma)$.

Theorem 1.6

Let $\Omega \subseteq R^N$ *be a domain satisfying simple*VL*-region*, $(a_{ij})=E$. $\lambda=\lambda_k$ $(k\geq 2)$. *Assume that* $F(x,s) \in C^1(\overline{\Omega} \times \mathbb{R}, \mathbb{R})$*.Furthermore, suppose the conditions* (F−1), (Fk−1), *and* (Fk−2). *Then the problem* (1.1) *has two nontrivial solutions in* $H^1_{p,\rho}(\Omega,\Gamma)$.

The paper is organized as follows. In Section 2, we provide and establish some lemmas which are necessary in the proof of our main theorems. In Section 3, we will prove Theorem 1.1-Theorem 1.5. In the last section, we establish Lemma 4.1 and Lemma 4.2, and we give the proof of Theorem 1.6.

2 PRELIMINARY RESULTS

In this section, we prove some lemmas which will be used in the proof of our main theorems. For simplicity, we denote $I_\lambda(\cdot)$ by $I(\cdot)$ in the following.

Lemma 2.1

Let L be defined as (1.2). *Then*

$$c_2\|u\|^2_{p,\rho} \geq L(u,u) \geq c_1\|u\|^2_{p,\rho}, \quad \forall u \in H^1_{p,\rho}(\Omega,\Gamma)$$

for some positive constants c_1, c_2.

To get Lemma 2.7 and Lemma 2.8, we first introduce a corollary of the Ekeland variation principle (see [16], Theorem 2.4).

Lemma 2.2

Let X be a Banach space. Assume that $E \in C^1(X,\mathbb{R})$ *is bounded from below, which satisfies the*(PS)c*condition. Then* $c = \inf_{x\in X} E(x)$ *is a critical value.*

Lemma 2.3

Assume that L is given by (1.2), *and the assumptions* (L−1)-(L−5) *hold, and that*(Ω,Γ) *is a*VL-*region. Then* $H^1_{p,\rho}(\Omega,\Gamma)$ *is compactly embedding in*L2ρ. *Moreover, if*(Ω,Γ) *is a simple*VL-*region and*(aij)=E, *then for* N≥2, $H^1_{p,\rho}(\Omega,\Gamma)$ *is compactly embedded in* $L^\theta_\rho, 2 < \theta < \frac{2N}{N-1}$.

Proof **See Lemma 2 and Theorem 9 in [12].** □

To establish the multiplicity of solutions for problem (1.1), we need to apply the following fundamental theorem (see [17], Theorem 9.12).

Lemma 2.4

Let X be an infinite-dimensional Banach space and let $E : X \to \mathbb{R}$ *be continuous, even and satisfying*(PS)c*for every*c∈R.*Assume, also, that*:

(1)There exist ρ>0, α>E(0), *and a subspace* V⊂X *of finite codimension such that*

$$\forall u \in V: \quad \|u\| = \rho \quad \Rightarrow \quad E(u) \geq \alpha.$$

(2)For every finite-dimensional subspace $W_n \subset X$, $\dim(W_n) = n$, *there exists* $R_n > 0$ *such that*

$$\forall u \in W_n: \quad \|u\| = R_n \quad \Rightarrow \quad E(u) \leq E(0).$$

Then there exists a sequence{hn}*of critical values of E with*hn→+∞.

Next, we describe some results under the geometry for functional *I*.

Lemma 2.5

Under hypotheses (F−1) *and* (F−2), (Ω,Γ) *is a*VL-*region, the functional I has the following saddle geometry for*λ=λ1:

(1) $I(u) \to \infty$ *if* $\|u\|_{p,\rho} \to \infty$ *with* $u \in V_1^{\perp}$.
(2) *There is* $\alpha \in \mathbb{R}$ *such that* $I(u) \leq \alpha$, $\forall u \in V_1$.
(3) $I(u) \geq b_1$, $\forall u \in V_1^{\perp}$, *where* b_1 *is given by* $(F-2)$.

Proof *(1)*By Remark 1.1, we have

$$I(u) \geq \frac{1}{2}\left(1 - \frac{\lambda_1}{\lambda_2}\right)L(u,u) + \int_{\Omega} F(x,u)\rho$$
$$\geq \frac{c_1}{2}\left(1 - \frac{\lambda_1}{\lambda_2}\right)\|u\|_{p,\rho}^2 + \int_{\Omega} F(x,u)\rho, \quad u \in V_1^{\perp}.$$

Using (F−1), we have I(u)→∞, as ‖u‖p,ρ→∞.

*(2)*By a simple calculation, we get

$$I(u) = \int_{\Omega} F(x,u), \quad u \in V_1.$$

By using (F−1) we have

$$I(u) = \int_{\Omega} F(x,u) \leq \int_{\Omega} k(x)\rho.$$

So we choose $\alpha = \int_{\Omega} k(x)\rho$.

*(3)*By (F−2) and Remark 1.1, we get

$$I(u) \geq \frac{1}{2}L(u,u) - \frac{\lambda_2}{2}\int_{\Omega} u^2\rho + b_1 \geq b_1, \quad \forall u \in V_1^{\perp},$$

the proof of this lemma is completed. □

Next, we will prove the Palais-Smale properties for the functional *I*. We recall that $I: E \to \mathbb{R}$ satisfies the Palais-Smale conditions at the level $c \in \mathbb{R}\,(PS)_c$ in short). For any sequence $\{h_n\} \subset E$ such that

$$I(h_n) \to c, \qquad I'(h_n) \to 0,$$

as n→∞, the sequence {hn} possesses a convergent subsequence in *E*. Moreover, we say that *I* satisfies (PS) conditions when we have (PS)c for all c∈R (see [16]).

Lemma 2.6

Assume the condition (F−1) *holds,* λ=λ1, *and*(Ω,Γ) *is a*VL-*region. Then the functional I has the*(PS)cconditions *whenever* $c < \min\{\int_{\Omega} T^+\rho, \int_{\Omega} T^-\rho\}$ *or* $c > \max\{\int_{\Omega} S^+\rho, \int_{\Omega} S^-\rho\}$.

Proof

We only prove the lemma for $c < \min\{\int_{\Omega} T^+\rho, \int_{\Omega} T^-\rho\}$. For the case $c > \max\{\int_{\Omega} S^+\rho, \int_{\Omega} S^-\rho\}$, we can use similar methods.

*1.*The boundedness of (PS) sequence.

If this is not the case, there exists a (PS)c unbounded sequence $\{u_n\} \in H^1_{p,\rho}(\Omega,\Gamma)$ such that $c < \min\{\int_{\Omega} T^+\rho, \int_{\Omega} T^-\rho\}$. Without loss of generality, we may assume that, as $n \to \infty$, the following expressions hold:

$$\|u_n\|_{p,\rho} \to \infty, \qquad I(u_n) \to c, \qquad I'(u_n) \to 0.$$

Define $\overline{u}_n = \frac{u_n}{\|u_n\|_{p,\rho}}$. Hence by Lemma 2.3, there is an $\overline{u} \in H^1_{p,\rho}(\Omega,\Gamma)$ with the following properties:

$\overline{u}_n \rightharpoonup \overline{u}$ in $H^1_{p,\rho}(\Omega,\Gamma)$,

$\overline{u}_n \rightharpoonup \overline{u}$ in $L^2_\rho(\Omega)$,

$\overline{u}_n \to \overline{u}$ a.e. in Ω.

For any $\Phi \in H^1_{p,\rho}(\Omega,\Gamma)$, it is obvious that $\frac{I'(u_n)\Phi}{\|u_n\|_{p,\rho}} \to 0$. By the convergence of $\{\overline{u}_n\}$ and (F−1), we have

$$L(\overline{u}, \Phi) = \lambda_1 \int_\Omega \overline{u}\Phi\rho.$$

According to the definition of λ_1, we obtain $\overline{u} = \pm\varphi_1$. So, we suppose initially that $\overline{u} = \varphi_1$. Because $\varphi_1 > 0$ it is obvious that $u_n \to +\infty$, a.e. $x \in \Omega$ as $n \to \infty$.

Hence, taking $u_n = t_n\varphi_1 + \omega_n$, $\{t_n\} \subset \mathbb{R}$, $\{\omega_n\} \subset V_1^\perp$, by Remark 1.1, we have

$$I(u_n) \geq \frac{1}{2}\left(1 - \frac{\lambda_1}{\lambda_2}\right) L(\omega_n, \omega_n) + \int_\Omega F(x, u_n)\rho. \tag{2.1}$$

Since $I(u_n) \to c$, it can easily be concluded that the sequence $\{\omega_n\}$ is bounded. Because of $\|u_n\| \to \infty$, on a subsequence $|t_n| \to \infty$, without loss of generality, we assume $t_n \to +\infty$.

Now, by the Hölder inequality and (F−1), we have

$$\left|\int_\Omega f(x,u_n)\omega_n\rho\right| \le C\left(\int_\Omega \left|f(x,u_n)\right|^2\rho\right)^{\frac{1}{2}}.$$

Thus, by applying the dominated convergence theorem, we conclude that

$$\lim_{n\to\infty}\int_\Omega f(x,u_n)\omega_n\rho = 0. \tag{2.2}$$

On the other hand,

$$I'(u_n)\omega_n = L(\omega_n,\omega_n) - \lambda_1\int_\Omega \omega_n^2\rho + \int_\Omega f(x,u_n)\omega_n\rho,$$

by (2.2) and Remark 1.1, we obtain

$$\left(1-\frac{\lambda_1}{\lambda_2}\right)L(\omega_n,\omega_n) \le \left|I'(u_n)\omega_n\right| + \left|\int_\Omega f(x,u_n)\omega_n\rho\right| \to 0,$$

as n→∞. Therefore, due to Remark 1.1, we have

$$L(\omega_n,\omega_n) - \lambda_1\int_\Omega \omega_n^2\rho \to 0, \quad \text{as } n\to\infty.$$

.

Consequently, by virtue of Fatou's lemma and assumption (F−1), we get

$$c = \lim_{n\to\infty}\left(L(\omega_n,\omega_n) - \lambda_1\int_\Omega \omega_n^2\rho + \int_\Omega F(x,u_n)\rho\right)$$
$$\ge \int_\Omega T^+\rho,$$

which contradicts with the condition $c < \min\{\int_\Omega T^+\rho, \int_\Omega T^-\rho\}$. Hence, the (PS)c sequence of the functional *I* is bounded.

*2.*Various convergences of {un}.

Since {un} is a bounded sequence, there is an $u \in H^1_{p,\rho}(\Omega,\Gamma)$ with the following properties:

un⇀u in $H^1_{p,\rho}(\Omega,\Gamma)$,

un→u in $L^2_\rho(\Omega)$,

un→u a.e. in Ω.

3.{un} converges to *u* in $H^1_{p,\rho}(\Omega,\Gamma)$.

From the definition of $(PS)_c$ sequence, we have, as n→∞,

$$I'(u_n)u = L(u_n, u) - \lambda_1 \int_\Omega u_n u\rho + \int_\Omega f(x, u_n)u\rho \to 0,$$
$$I'(u_n)u_n = L(u_n, u_n) - \lambda_1 \int_\Omega u_n^2\rho + \int_\Omega f(x, u_n)u_n\rho \to 0.$$

By Fatou's lemma and the above convergence of {un}, it is easy to show that

$$\int_\Omega u_n u\rho \to \int_\Omega u^2\rho,$$
$$\int_\Omega f(x, u_n)u\rho \to \int_\Omega f(x, u)u\rho,$$
$$\int_\Omega u_n^2\rho \to \int_\Omega u^2\rho,$$
$$\int_\Omega f(x, u_n)u_n\rho \to \int_\Omega f(x, u)u\rho,$$

as n→∞. Hence, we get

$$L(u_n, u) \to \lambda_1 \int_\Omega u^2\rho - \int_\Omega f(x, u)u\rho, \quad \text{as } n \to \infty, \tag{2.3}$$

$$L(u_n, u_n) \to \lambda_1 \int_\Omega u^2 \rho - \int_\Omega f(x,u)u\rho, \quad \text{as } n \to \infty. \tag{2.4}$$

By the weak convergence, it follows that

$$L(u_n, u) \to L(u,u), \quad \text{as } n \to \infty. \tag{2.5}$$

By using (2.3), (2.4), (2.5), and a simple calculation, we obtain

$$L(u_n - u, u_n - u) \to 0, \quad \text{as } n \to \infty.$$

From Lemma 2.1, the proof is completed. □

Lemma 2.7

Suppose that (F−1) *and* (F−6) *are satisfied,* $\lambda=\lambda_1$ *and* (Ω,Γ) *is a simple*VL*-region. Then the origin is a local minimum for the functional* *I.*

Proof

From (1.4), it is easy to see that $I(0)=0$. By (F−6), we can choose $q \in (2, \frac{2N}{N-1})$ and a constant $C > 0$ such that

$$F(x,s) \geq \frac{1-\alpha}{2}\lambda_1 s^2 - C|s|^q, \quad \forall (x,s) \in \Omega \times \mathbb{R}.$$

Consequently, by Remark 1.1 and Lemma 2.3, we have

$$I(u) \geq \frac{1}{2}(1-\alpha)c_1\|u\|^2_{p,\rho} - C\|u\|^q_{p,\rho} \geq \frac{1}{4}(1-\alpha)c_1\|u\|^2_{p,\rho}, \quad \|u\|_{p,\rho} < r,$$

where *r* is small enough and $0 < r < t_0$, t_0 is provided by (F−4). Therefore the proof is completed. □

To complete the mountain-pass geometry, we also need the following result.

Lemma 2.8

Let the hypotheses (F−1), (F−3), (F−4), *and* (F−6) *hold,* λ=λ1*and let*(Ω,Γ) *be a*VL*-region. Then there exists* $u_0 \in H^1_{p,\rho}(\Omega,\Gamma)$ *such that* $I(u_0) < 0$ *and* $\|u_0\|_{p,\rho} > r$*, where r is given by Lemma* 2.7.

Proof

By (F−1) and (F−4), we take u0=t0φ1 where t0 is provided by (F−4). Thus, we obtain

$$I(t_0\varphi_1) < \min\left\{\int_\Omega T^+\rho, \int_\Omega T^-\rho\right\} < \max\left\{\int_\Omega S^+\rho, \int_\Omega S^-\rho\right\} \leq 0,$$

and $\|t_0\Phi_1\| = t_0$. By Lemma 2.7, we have 0<r<t0, then the conclusion follows. □

Lemma 2.9

If hypotheses (F−1), (F−3), *and* (F−4) *are satisfied,* λ=λ1*and*(Ω,Γ) *is a* V_L*-region. Then problem* (1.1) *has at least one nontrivial solution* $u_0 \in H^1_{p,\rho}(\Omega,\Gamma)$. *Moreover,* u0 u_0 *has negative energy, i.e.* $I(u_0) < 0$.

Proof

By (F−1) and Remark 1.1, we have

$$I(u) \geq \int_\Omega F(x,u)\rho \geq -\int_\Omega k(x)\rho.$$

Therefore, the functional *I* is bounded below. In this case, we would like to mention that the functional *I* satisfies the (PS)c conditions with $c = \inf\{I(h) : h \in H^1_{p,\rho}(\Omega,\Gamma)\}$ To see this, by Lemma 2.6, we take t0∈R provided by (F−4), we obtain

$$c \leq I(t_0\varphi_1) = \int_\Omega F(x, t_0\varphi_1) < \min\left\{\int_\Omega T^+\rho, \int_\Omega T^-\rho\right\} \leq 0.$$

Consequently, applying Lemma 2.2 we have one critical point $u_0 \in H^1_{p,\rho}(\Omega,\Gamma)$ such that $I(u_0) = \inf\{I(h) : h \in H^1_{p,\rho}(\Omega,\Gamma)\} \leq I(t_0\varphi_1) < 0$. The proof of this lemma is completed. □

To prove Theorem 1.3, we establish the following lemma.

Lemma 2.10

Assume that the conditions (F−1), (F−2), (F−3), *and* (F−5) *are satisfied,* λ=λ1*and*(Ω,Γ) *is a*VL-*region. Then the problem* (1.1) *has at least two nontrivial solutions with negative energy.*

Proof Define

$$M^+ = \{t\varphi_1 + \omega, t \geq 0, \omega \in V_1^\perp\}, \qquad M^- = \{t\varphi_1 + \omega, t \leq 0, \omega \in V_1^\perp\}.$$

We have $\partial M^+ = \partial M^- = V_1^\perp$. Hence, we minimizer the functional *I* restrict to M+ and M−.

Firstly, we consider the functionals I±=I|M±. By the assumption (F−4), we have

$$c^\pm = \inf\{I^\pm, h \in E\} \leq I^\pm(t_1^\pm\varphi_1) = \int_\Omega F(x, t_1^\pm\varphi_1)\rho$$

$$< \min\left\{\int_\Omega T^+\rho, \int_\Omega T^-\rho\right\} \leq 0.$$

By Lemma 2.6, I± satisfy the (PS)c conditions whenever $c < \min\{\int_\Omega T^+\rho, \int_\Omega T^-\rho\}$. Therefore, we find that I± satisfy the (PS)c conditions with $c^+ = \inf\{I^+(h) : h \in E\}$.

In this way, by using Lemma 2.2 for the functionals I+ and I−, we obtain two critical points denoted by u_0^+ and u_0^-, respectively. Thus, we have $c^+ = I^+(u_0^+) = \inf_{h\in M^+}\{I(h)\}$ and $c^- = I^-(u_0^-) = \inf_{h\in M^-}\{I(h)\}$.

Moreover, we affirm that u_0^+ and u_0^- are nonzero critical points. Based on (F−3) and (F−5), we obtain

$$I^{\pm}(u_0^{\pm}) \leq I^{\pm}(t_1^{\pm}\varphi_1) < \min\left\{\int_{\Omega} T^+\rho, \int_{\Omega} T^-\rho\right\} \leq 0,$$

and I is restricted to V⊥1 being nonnegative. More specifically, given ω∈V⊥1 and (3) in Lemma 2.5, we have

$$I(\omega) \geq b_1 \geq 0. \tag{2.6}$$

Next, we prove that u+0 and u−0 are distinct. The proof of this affirmation is by contradiction. If $u_0^+ = u_0^-$, then $u_0^+ = u_0^- \in V_1^{\perp}$. By (2.6), we obtain $I(u_0^+) < 0 \leq I(u_0^+)$. Therefore, we have a contradiction. Consequently, we get $u_0^+ \neq u_0^-$. Thus the problem (1.1) has at least two nontrivial solutions. Moreover, these solutions have negative energy.

To prove Theorem 1.5, we need the following lemma. The proof is similar to that of Lemma 2.6.

Lemma 2.11

Assume that the conditions (F′−1) *and* (F′−2) *hold,* λ<λ1, (aij)=E*and*(Ω,Γ) *is a simple*VL*-region. Then functional Isatisfies the*(PS)*cconditions.*

3 PROOFS OF THEOREM 1.1-THEOREM 1.5

In this section, we prove Theorems 1.1, 1.2, 1.3, 1.4, and 1.5.

Proof of Theorem 1.1 By Lemma 2.9, we get a solution u0 which satisfies I(u0)<0. It follows that the problem (1.1) has at least one nontrivial solution. The proof is completed. □

Proof of Theorem 1.2 Using Lemma 2.10, we obtain two distinct critical points u_0^+ and u_0^- such that $I(u_0^+) < 0$ and $I(u_0^-) < 0$. Therefore, the problem (1.1) has at least two nontrivial solutions. The proof is completed. □

Proof of Theorem 1.3

From Lemma 2.7 and Lemma 2.8, we know that the functional *I* satisfies the geometric conditions of the mountain-pass theorem. Moreover, the functional *I* satisfies the (PS)c conditions for all c≥0. Thus, we have a solution $u_2 \in H^1_{p,\rho}(\Omega, \Gamma)$ given by the mountain-pass theorem. Obviously, the solution u2 satisfies I(u2)>0.

On the other hand, by Lemma 2.9, we get another solution u0 satisfying I(u0)<0. It follows that the problem (1.1) has at least two nontrivial solutions. □

Proof of Theorem 1.4

The conditions (F−1), (F−3), (F−5), and (F−6) imply that Lemma 2.7 and Lemma 2.8 hold. Thus, we have one solution u2 which satisfies $I(u_2) > 0$.

On the other hand, using Lemma 2.10, we obtain two distinct critical points $u_0^\pm$ such that I $u_0^\pm$ <0. In other words, the problem (1.1) has at least three nontrivial solutions. The proof is completed. □

Proof of Theorem 1.5

It is easy to see that functional *I* is continuous and even. Moreover, by Lemma 2.11, *I* satisfies (PS)c condition for every c∈R. To prove Theorem 1.6, we only need to test and verify the conditions (1) and (2) in Lemma 2.4.

*1.*By (F′−1) and Remark 1.1, we have

$$I(u) \geq \frac{1}{2}\left(1-\frac{\lambda}{\lambda_1}\right)c_1\|u\|_{p,\rho}^2 - C\|u\|_{p,\rho}^{\theta+1}.$$

Hence there exist r>0 small enough and δ>0 such that I(u)≥δ, for $\|u\|_{p,\rho}=r.$. That is to say the condition (1) in Lemma 2.4 holds with $V=H_{p,\rho}^1(\Omega,\Gamma)$.

*2.*We verify the condition (2) in Lemma 2.4. Let *W* be a finite-dimensional subspace of $H_{p,\rho}^1(\Omega,\Gamma)$. Let u∈W such that I(u)≥0, *i.e.*

$$\frac{1}{2}L(u,u)-\frac{\lambda}{2}\int_\Omega u^2\rho+\int_\Omega F(x,u)\rho \geq 0. \tag{3.1}$$

By (F′−1), there exist m(x)∈L∞(Ω) satisfying m(x)>0 a.e. x∈Ω and a positive constant *c* such that

$$F(x,s) \leq -m(x)|s|^\eta + cs^2, \quad \text{a.e. } x\in\Omega \text{ and } \forall s\in\mathbb{R}.$$

The inequality (3.1) implies

$$c_2\|u\|_{p,\rho}^2 \geq \int_\Omega m(x)|u|^\eta\rho - c\int_\Omega u^2\rho. \tag{3.2}$$

Since $(\int_\Omega m(x)|u|^\eta)^{\frac{1}{\eta}}$ is a norm on *W* and *W* is finite-dimensional, then, by (3.2), there exists c>0 such that $\|u\|_{p,\rho}^\eta \leq c\|u\|_{p,\rho}^2$. Since η>2, we deduce that the set {u∈W,I(u)≥0} is bounded in $H_{p,\rho}^1(\Omega,\Gamma)$ and the condition (2) in Lemma 2.4 holds. □

4 PROOF OF THEOREM 1.6

In this section, we consider the problem (1.1) in λ=λk. In order to prove Theorem 1.6, we first establish the following lemmas.

Lemma 4.1

Assume that the condition (F−1) *holds, and that*(Ω,Γ) *is a*VL*-region. Then the functional I satisfies the*(PS)c *conditions whenever* $c < \inf_{\omega\in V(\lambda_k)}\{\int_{\Omega\cap\{\omega>0\}} T^+\rho + \int_{\Omega\cap\{\omega<0\}} T^-\rho\}$ *or* $c > \sup_{\omega\in V(\lambda_k)}\{\int_{\Omega\cap\{\omega>0\}} S^+\rho + \int_{\Omega\cap\{\omega<0\}} S^-\rho\}$.

Proof We only prove the lemma for all $c < \inf_{\omega\in V(\lambda_k)}\{\int_{\Omega\cap\{\omega>0\}} T^+\rho + \int_{\Omega\cap\{\omega<0\}} T^-\rho\}$. For the case $c > \sup_{\omega\in V(\lambda_k)}\{\int_{\Omega\cap\{\omega>0\}} S^+\rho + \int_{\Omega\cap\{\omega<0\}} S^-\rho\}$, we can use similar methods.

The boundedness of (PS)c sequence.

Let us prove by contradiction. Suppose that there exists a (PS)c unbounded sequence $\{u_n\}\in H^1_{p,\rho}(\Omega,\Gamma)$ such that $c < \inf_{\omega\in V(\lambda_k)}\{\int_{\Omega\cap\{\omega>0\}} T^+\rho + \int_{\Omega\cap\{\omega<0\}} T^-\rho\}$. For ease of notation and without loss of generality, we assume that

$$\|u_n\|_{p,\rho}\to\infty,\qquad I(u_n)\to c,\qquad I'(u_n)\to 0,\quad n\to\infty.$$

Define $\overline{u}_n = \frac{u_n}{\|u_n\|_{p,\rho}}$. Hence by Lemma 2.3, there is an $\overline{u}\in H^1_{p,\rho}(\Omega,\Gamma)$ with the following properties:

$$\overline{u}_n \rightharpoonup \overline{u}\quad \text{in } H^1_{p,\rho}(\Omega,\Gamma),\qquad \overline{u}_n\to\overline{u}\quad \text{in } L^2_\rho(\Omega),\qquad \overline{u}_n\to\overline{u},\quad \text{a.e. in } \Omega.$$

For any $\Phi\in H^1_{p,\rho}(\Omega,\Gamma)$, we have $\frac{I'(u_n)\Phi}{\|u_n\|_{p,\rho}}\to 0.$ By the convergence of {u¯n} and (F−1), we have

$$L(\overline{u},\Phi)=\lambda_k\int_\Omega \overline{u}\Phi\rho.$$

By the definition of λk, we obtain u⁻∈V(λk). It is obvious that un(x)→+∞, x∈{x∈Ω|u⁻(x)>0} as n→∞, and un(x)→−∞, x∈{x∈Ω|u⁻(x)<0} as n→∞. By (VL−1), we have m({x∈Ω|u⁻(x)=0})=0.

Hence, we can take $u_n = \omega_n^{(1)} + \varphi_n + \omega_n^{(2)}$, where $\omega_n^{(1)} \in V_1$, $\varphi_n \in V(\lambda_k)$, $\omega_n^{(2)} \in V_2$. $L_\rho^2 = V_1 \oplus V(\lambda_k) \oplus V_2$, V_1, V1denotes the eigenspace associated to the eigenvalue λ<λk and V2 denotes the eigenspace associated to the eigenvalue λ>λk. Set

$$\widetilde{u}_n = -\omega_n^{(1)} + \omega_n^{(2)}.$$

By Remark 1.1, we have

$$I'(u_n)\widetilde{u}_n \geq \min\left\{\frac{\lambda_k}{\lambda_{k-1}} - 1, 1 - \frac{\lambda_k}{\lambda_{k+1}}\right\} L(u_n - \varphi_n, u_n - \varphi_n) + \int_\Omega f(x, u_n)\widetilde{u}_n \rho.$$

Hence, by Lemma 2.1, we get

$$L(u_n - \varphi_n, u_n - \varphi_n) \leq C\|\widetilde{u}_n\|_{p,\rho} \leq C \cdot L^{\frac{1}{2}}(u_n - \varphi_n, u_n - \varphi_n),$$

and $u_n - \varphi_n$ is bounded in $H^1_{p,\rho}(\Omega, \Gamma)$, *i.e.*, $\omega_n^{(1)} + \omega_n^{(2)}$ is bounded in $H^1_{p,\rho}(\Omega, \Gamma)$. Letting $\omega_n = \omega_n^{(1)} + \omega_n^{(2)}$, one has

$$I'(u_n)\omega_n = L(\omega_n, \omega_n) - \lambda_k \int_\Omega \omega_n^2 \rho + \int_\Omega f(x, u_n)\omega_n \rho.$$

By (F−1), the Hölder inequality, and the Lebesgue dominated theorem, it is easy to show

$$\int_\Omega f(x, u_n)\omega_n \rho \to 0, \quad \text{as } n \to \infty.$$

Since I′(un)→0 and ωn is bounded in $H^1_{p,\rho}(\Omega, \Gamma)$, we have I′(un)ωn→0 and

$$L(\omega_n, \omega_n) - \lambda_k \int_\Omega \omega_n^2 \rho \to 0, \quad \text{as } n \to \infty. \tag{4.1}$$

Consequently, by virtue of Fatou's lemma, (4.1) and (F−1), we have

$$c \geq \int_\Omega \liminf_{n\to\infty} F(x, u_n)\rho = \int_{\Omega\cap\{\bar{u}>0\}} T^+\rho + \int_{\Omega\cap\{\bar{u}<0\}} T^-\rho,$$

which contradicts with the condition $c < \inf_{\omega\in V(\lambda_k)}\{\int_{\Omega\cap\{\omega>0\}} T^+\rho + \int_{\Omega\cap\{\omega<0\}} T^-\rho\}$. Hence, the (PS)c sequence of the functional *I* is bounded.

We can argue as the proof of Lemma 2.6, so the proof has been completed. □

Lemma 4.2

Suppose that (F−1) *and* (Fk−2) *are satisfied,* (Ω,Γ) *is a simple*VL-*region and*(aij)=E. *Then the origin is a local minimum for the functional I.*

Proof

By (Fk−2), we can choose $q \in (2, \frac{2N}{N-1})$ and a constant C>0 such that

$$F(x, s) \geq \gamma s^2 - C|s|^q, \quad \forall (x, s) \in \Omega \times \mathbb{R}.$$

Consequently, by Remark 1.1 and Lemma 2.3, we have

$$I(u) \geq \frac{1}{4}\left(1 + \frac{2\gamma}{\lambda_1} - \frac{\lambda_k}{\lambda_1}\right) c_1 \|u\|_{p,\rho}^2,$$

$\|u\|_{p,\rho}<r$, where *r* is small enough. Therefore the proof is completed.

Proof of Theorem 1.6 Since

$$I(t\varphi_1) \leq (\lambda_1 - \lambda_k)\frac{t^2}{2} + \int_\Omega k\rho \to -\infty,$$

as $t \to \infty$, we can choose $t_+>0$ and $t_-<0$, satisfying $\|t_+\varphi_1\|_{p,\rho}>r$ and $\|t_-\varphi_1\|_{p,\rho}>r$ such that $I(t_+\varphi_1)<0$ and $I(t_-\varphi_1)<0$, where *r* is given by Lemma 4.2.

On the other hand, since $\varphi_1>0$, we know $t_+\varphi_1>0$ and $t_-\varphi_1<0$. Let $P_+ = \{u \in H^1_{p,\varphi}(\Omega,\Gamma) \mid u \geq 0\}$ and $P_- = \{u \in H^1_{p,\varphi}(\Omega,\Gamma) \mid u \leq 0\}$. Consider the functionals $I_\pm$, which are the restrictions of *I* on $P_\pm$. By Lemma 4.1 and (F_{k-1}), for $c \geq 0$, we get $I_\pm$ satisfying $(PS)_c$ conditions. So by the mountain-pass lemma, we get two critical points $z_\pm \in P_\pm$. It is obvious that $z_\pm \neq 0$. The proof has been completed. □

ACKNOWLEDGEMENTS

The authors express their sincere thanks to the referees for their valuable criticisms of the manuscript and for helpful suggestions. This work was supported by National Natural Science Foundation of China (11171220) and Shanghai Leading Academic Discipline Project (XTKX2012).

COMPETING INTERESTS

The authors declare that they have no competing interests.

AUTHORS' CONTRIBUTIONS

All authors collaborated and dedicated the same amount of time in working on this article. All authors read and approved the final manuscript.

REFERENCES

1. Ahmad S, Lazer AC, Paul JL: Elementary critical point theory and perturbations of elliptic boundary value problems at resonance. *Indiana Univ. Math. J.* 1976, 25: 933-944.

2. Bartsch T, Chang KC, Wang ZQ: On the Morse indices of sign changing solutions of nonlinear elliptic problems. *Math. Z.* 2000, 233: 655-677.

3. Bartsch T, Li SJ: Critical point theory for asymptotically quadratic functionals and applications to problems with resonance. *Nonlinear Anal.*1997, 28: 419-441.

4. Chang KC: An extension of the Hess-Kato theorem to elliptic systems and its applications to multiple solutions problems. *Acta Math. Sin.*1999, 15: 439-454.

5. Chang KC: *Infinite Dimensional Morse Theory and Multiple Solution Problems*. Birkhäuser, Boston; 1993.

6. Chang KC: Principal eigenvalue for weight in elliptic systems. *Nonlinear Anal.* 2001, 46: 419-433.

7. Costa DG, Magalhes CA: A variational approach to subquadratic perturbations of elliptic systems. *J. Differ. Equ.* 1994, 111: 103-121.

8. de Figueiredo DG: Positive solutions of semilinear elliptic problems. In *Differential Equations*. Springer, Berlin; 1982:34-87. (São Paulo, 1981)

9. Furtado MF, de Paiva FOV: Multiplicity of solutions for resonant elliptic systems. *J. Math. Anal. Appl.* 2006, 319: 435-449.

10. Landesman EM, Lazer AC: Nonlinear perturbations of linear elliptic boundary value problems at resonance. *J. Math. Mech.* 1970, 19: 609-623.

11. Pao CV: *Nonlinear Parabolic and Elliptic Equations*. Plenum, New York; 1992.

12. Shapiro VL: Singular quasilinearity and higher eigenvalues. *Mem. Am. Math. Soc.* 2001., 153:

13. Jia G, Zhao Q: Existence results in weighted Sobolev spaces for some singular quasilinear elliptic equations. *Acta Appl. Math.* 2010, 109: 599-603.

14. Jia G, Sun D: Existence of solutions for a class of singular quasilinear elliptic resonance problems. *Nonlinear Anal.* 2011, 74: 3055-3064.

15. Jia G, Huang LN, Zhang XJ: Existence of solutions for quasilinear elliptic equations with superlinear nonlinearities. *Bound. Value Probl.*2012., 2012:

16. Willem M: *Minimax Theorems*. Birkhäuser, Boston; 1996.

17. Rabinowitz PH: On a class of nonlinear Schrödinger equations. *Z. Angew. Math. Phys.* 1992, 43: 270-291.

10

EXISTENCE RESULT FOR SEMILINEAR ELLIPTIC SYSTEMS INVOLVING CRITICAL EXPONENTS

S Khademloo*, M Farzinejad and O Khazaee kohpar

Department of Basic Sciences, Babol University of Technology, Babol, 47148-71167, Iran

ABSTRACT

In this paper we deal with the existence of a positive solution for a class of semilinear systems of multi-singular elliptic equations which involve Sobolev critical exponents. In fact, by the analytic techniques and variational methods, we prove that there exists at least one positive solution for the system.

KEYWORDS

semilinear elliptic system, nontrivial solution, critical exponent variational method

1 INTRODUCTION

We consider the following elliptic system:

$$\begin{cases} Lu = \frac{\sigma\alpha}{2^*}u|u|^{\alpha-2}|v|^{\beta} + \eta u|u|^{2^*-2} + a_1u + a_2v, & \Omega, \\ Lv = \frac{\sigma\beta}{2^*}v|v|^{\beta-2}|u|^{\alpha} + \lambda v|v|^{2^*-2} + a_2u + a_3v, & \Omega, \\ u = v = 0, & \partial\Omega, \end{cases} \tag{1.1}$$

where Ω⊂RN (N≥3) is a smooth bounded domain such that $\xi_i \in \Omega, i = 1,2,\ldots,k, k \geq 2,$ are different points,

$0 \leq \mu_i < \bar{\mu} := (\frac{N-2}{2})^2, L := -\Delta \cdot - \sum_{i=1}^{k} \mu_i \frac{\cdot}{|x-\xi_i|^2}, \eta, \lambda, \sigma \geq 0, a_1, a_2, a_3 \in$.

$\mathbb{R}, 1 < \alpha, \beta < 2^* - 1, \alpha + \beta = 2^*.$

We work in the product space H×H, where the space $H := H_0^1(\Omega)$ is the completion of $C_0^\infty(\Omega)$ with respect to the norm $(\int_\Omega |\nabla \cdot|^2 dx)^{\frac{1}{2}}$.

In resent years many publications [1–3] concerning semilinear elliptic equations involving singular points and the critical Sobolev exponent have appeared. Particularly in the last decade or so, many authors used the variational method and analytic techniques to study the existence of positive solutions of systems of the form of (1.1) or its variations; see, for example, [4–8].

Before stating the main result, we clarify some terminology. Since our method is variational in nature, we need to define the energy functional of (1.1) on H×H

$$\begin{aligned} J(u,v) = &\frac{1}{2}\int_\Omega \left(|\nabla u|^2 + |\nabla v|^2 - \sum_{i=1}^{k} \frac{\mu_i(u^2+v^2)}{|x-\xi_i|^2} \right) dx \\ &- \frac{1}{2}\int_\Omega \left(a_1u^2 + 2a_2uv + a_3v^2\right) dx - \sigma\frac{\sigma}{2^*}\int_\Omega |u|^{\alpha}|u|^{\beta}\,dx \\ &- \frac{\eta}{2^*}\int_\Omega |u|^{2^*}dx - \frac{\lambda}{2^*}\int_\Omega |v|^{2^*}dx. \end{aligned}$$

Then J(u,v) belongs to C^1(H×H,R). A pair of functions (u0,v0)∈H×H is said to be a solution of (1.1) if (u0,v0)≠(0,0), and for all (φ,ϕ)∈H×H, we have

$$\int_{\Omega}\left(\nabla u_0\cdot\nabla\varphi+\nabla v_0\cdot\nabla\phi-\sum_{i=1}^{k}\frac{\mu_i(u_0\varphi+v_0\phi)}{|x-\xi_i|^2}-(a_1u_0\varphi+a_2\varphi v_0+a_2\phi u_0+a_3v_0\phi)\right.$$
$$-\sigma\left(\alpha\varphi u_0|u_0|^{\alpha-2}|v_0|^{\beta}+\beta\phi v_0|v_0|^{\beta-2}|u_0|^{\alpha}\right)$$
$$\left.-\eta|u_0|^{2^*-2}u_0\varphi-\lambda|v_0|^{2^*-2}v_0\phi\right)dx=\langle J'(u_0,v_0),(\varphi,\phi)\rangle=0.$$

Standard elliptic arguments show that

$$u,v\in C^2\left(\Omega\backslash\{\xi_1,\ldots,\xi_k\}\right)\cap C^1\left(\overline{\Omega}\backslash\{\xi_1,\ldots,\xi_k\}\right).$$

The following assumptions are needed:

$(\mathcal{H}_1)$ $\eta+\lambda+\sigma>0$, $0\le\mu_1\le\mu_2\le\cdots\le\mu_k<\bar{\mu}-1$ and $\sum_{i=1}^{k}\mu_i<\bar{\mu}$, $\alpha+\beta>1$, $\alpha+\beta=2^*$, $a_1,a_2,a_3>0$, $a_1a_3-a_2^2>0$,

($(\mathcal{H}_2)$ $0<\lambda_1\le\lambda_2<\Lambda_1(\mu)$, where $\Lambda1(\mu)$ is the first eigenvalue of L, $\lambda1$, $\lambda2$ are the eigenvalues of the matrix $A=\begin{pmatrix}a_1 & a_2\\ a_2 & a_3\end{pmatrix}$.

The quadratic from $Q(u,v):=(u,v)A(u,v)^T=a_1u^2+2a_2uv+a_3v^2$ is positively defined and satisfies

$$\lambda_1\left(u^2+v^2\right)\le a_1u^2+2a_2uv+a_3v^2\le\lambda_2\left(u^2+v^2\right)\quad\forall u,v\in H. \tag{1.2}$$

Our main results are as follows.

Theorem 1.1 *Suppose* (H1) *holds. Then for any solution* (u,v)∈H×H *of problem* (1.1), *there exists a positive constant* C1*such that*

$$\max\left\{|u(x)|,|v(x)|\right\}\le C_1|x-\xi_i|^{-(\sqrt{\bar{\mu}}-\sqrt{\bar{\mu}-\mu_i})},\quad\forall x\in B_{\rho_1}(\xi_i)\backslash\{\xi_i\},$$

where ρ1>0 *and* Bρ1(ξi)⊂Ω.

Theorem 1.2 *Suppose* (H1) *holds. Then for any positive solution* (u,v)∈H×H *of problem* (1.1), *there exists a positive constant* C2C_2 *such that* Bρ2(ξi)⊂Ω *and*

$$\min\{u(x), v(x)\} \geq C_2|x-\xi_i|^{-(\sqrt{\bar{\mu}}-\sqrt{\bar{\mu}-\mu_i})}, \quad \forall x \in B_{\rho_2}(\xi_i)\backslash\{\xi_i\},$$

where ρ2>0.

Theorem 1.3 *Suppose* (H1), (H2) *hold. Then the problem* (1.1) *has a positive solution.*

2 PRELIMINARIES

On H×H, we use the norm

$$\|(u,v)\|_{H\times H} = \left(\int_\Omega (|\nabla u|^2 + |\nabla v|^2)\,dx\right)^{\frac{1}{2}}.$$

Using the Young inequality, the following best constant is well defined:

$$S_{\mu_i} := \inf_{u\in D^{1,2}(\mathbb{R}^N)\backslash\{0\}} \frac{\int_{\mathbb{R}^N}(|\nabla u|^2 - \mu_i \frac{u^2}{|x-\xi_i|^2})\,dx}{(\int_{\mathbb{R}^N}|u|^{2^*}\,dx)^{2/2^*}}, \tag{2.1}$$

where D1,2(RN) is the completion of $C_0^\infty(\mathbb{R}^N)$ with respect to the norm $(\int_{\mathbb{R}^N}|\nabla\cdot|^2\,dx)^{\frac{1}{2}}$.

We infer that Sμi is attained in R^N by the functions

$$V_{\mu_i,\varepsilon}^{\xi_i}(x) = \varepsilon^{\frac{2-N}{2}} U_{\mu_i}\left(\frac{x-\xi_i}{\varepsilon}\right), \quad \forall\varepsilon > 0,$$

where

$$U_{\mu_i}(x-\xi_i)=\frac{\left(\frac{2N(\bar{\mu}-\mu_i)}{\sqrt{\bar{\mu}}}\right)^{\sqrt{\bar{\mu}}/2}}{|x-\xi_i|^{(\sqrt{\bar{\mu}}-\sqrt{\bar{\mu}-\mu_i})}\left(1+|x-\xi_i|^{\frac{2\sqrt{\bar{\mu}-\mu_i}}{\sqrt{\bar{\mu}}}}\right)^{\sqrt{\bar{\mu}}}}.$$

For all η,λ,σ≥0, η+λ+σ>0, α+β>1, α+β=2∗, by the Young and Hardy-Sobolev inequalities, the following constant is well defined on $\mathcal{D}:=(D^{1,2}(\mathbb{R}^N)\backslash\{0\})^2$:

$$S_{\eta,\lambda,\sigma}(\mu_i):=\inf_{(u,v)\in\mathcal{D}}\frac{\int_{\mathbb{R}^N}(|\nabla u|^2+|\nabla v|^2-\mu_i\frac{u^2+v^2}{|x-\xi_i|^2})\,dx}{(\int_{\mathbb{R}^N}(\eta|u|^{2^*}+\lambda|v|^{2^*}+\sigma|u|^{\alpha}|v|^{\beta})\,dx)^{2/2^*}}. \tag{2.2}$$

Set

$$u^{\xi}_{\mu,\varepsilon}(x)=\psi(x)V^{\xi}_{\mu,\varepsilon}(x)=\varepsilon^{\frac{2-N}{2}}\psi(x)U_{\mu}\left(\frac{x-\xi}{\varepsilon}\right),$$

where ξ∈Ω, 0≤μ<μ¯, ψ∈C∞0(Bρ(ξ)) satisfies 0≤ψ≤1 and ψ≡1, ∀x∈Bρ2(ξ), for all ρ>0 small. Then for any 0<μ<μ¯, by [9] we have the following estimates:

$$\int_{\Omega}\left(|\nabla u^{\xi}_{\mu,\varepsilon}|^2-\mu\frac{(u^{\xi}_{\mu,\varepsilon})^2}{|x-\xi|^2}\right)dx=S_{\mu}^{\frac{N}{2}}+o(\varepsilon^{2\sqrt{\bar{\mu}-\mu}}),$$

$$\int_{\Omega}|u^{\xi}_{\mu,\varepsilon}|^{2^*}dx=S_{\mu}^{\frac{N}{2}}+o(\varepsilon^{\frac{2N}{N-2}\sqrt{\bar{\mu}-\mu}}),$$

and for any $a\in\mathbb{R}^N\backslash\{0\}$,

$$\int_{\Omega}\frac{|u^0_{\mu,\varepsilon}|^2}{|x+\xi|^2}dx=\begin{cases}\frac{\varepsilon^2}{|\xi|^2}\int_{\mathbb{R}^N}U^2_{\mu}(x)\,dx+o(\varepsilon^2) & \text{if } \mu<\bar{\mu}-1,\\ \frac{C^2_{\mu}\omega_N}{|\xi|^2}\varepsilon^2|\log\varepsilon|+o(\varepsilon^2) & \text{if } \mu=\bar{\mu}-1,\end{cases}$$

$$\int_{\Omega}|u^{\xi}_{\mu,\varepsilon}|^2dx=\begin{cases}o_1(\varepsilon^2) & \text{if } 0\le\mu<\bar{\mu}-1,\\ o_1(\varepsilon^2|\log\varepsilon|) & \text{if } \mu=\bar{\mu}-1,\end{cases}$$

where $C_{\mu}=(\frac{4N(\bar{\mu}-\mu)}{N-2})^{\frac{N-2}{4}}$, ωN$\omega_N$ is the volume of the unit ball in R^N.

3 ASYMPTOTIC BEHAVIOR OF SOLUTIONS

Proof of Theorem 1.1 Suppose (u0,v0)∈H×H is a nontrivial solution to problem (1.1). For all 0≤μi≤μ¯ define

$$u(x) = |x-\xi_i|^{\gamma} u_0(x) \quad \text{and} \quad \nu(x) = |x-\xi_i|^{\gamma} \nu_0(x), \quad \text{where } \gamma = (\sqrt{\bar{\mu}} - \sqrt{\bar{\mu}-\mu_i}).$$

It is not difficult to verify that $u, \nu \in H_0^1(\Omega, |x-\xi_i|^{-2\gamma})$ and satisfy

$$\begin{cases} -\operatorname{div}(|x-\xi_i|^{-2\gamma}\nabla u) \\ \quad = \frac{\sigma\alpha}{2^*}|x-\xi_i|^{-2^*\gamma} u|u|^{\alpha-2}|\nu|^{\beta} + \eta|x-\xi_i|^{-2^*\gamma} u|u|^{2^*-2} \\ \qquad + a_1|x-\xi_i|^{-2\gamma}u + a_2|x-\xi_i|^{-2\gamma}\nu + \sum_{\substack{j=1\\ j\neq i}}^{k} \frac{\mu_j}{|x-\xi_j|^2}|x-\xi_i|^{-2\gamma}u, \\ -\operatorname{div}(|x-\xi_i|^{-2\gamma}\nabla \nu) \\ \quad = \frac{\sigma\beta}{2^*}|x-\xi_i|^{-2^*\gamma} \nu|\nu|^{\beta-2}|u|^{\alpha} + \lambda|x-\xi_i|^{-2^*\gamma} \nu|\nu|^{2^*-2} \\ \qquad + a_2|x-\xi_i|^{-2\gamma}u + a_3|x-\xi_i|^{-2\gamma}\nu + \sum_{\substack{j=1\\ j\neq i}}^{k} \frac{\mu_j}{|x-\xi_j|^2}|x-\xi_i|^{-2\gamma}\nu. \end{cases} \tag{3.1}$$

Let R>0 small enough such that BR(ξi)⊂Ω and ξi∉BR(ξi) for j≠i$j \neq i$. Also, let φi∈$C_0^{\infty}(B_R(\xi_i))$ be a cut-off function. Set

$$u_n := \min\{|u|, n\}; \qquad \nu_n := \min\{|\nu|, n\}; \qquad \phi_{1_i} := \varphi_i^2 u u_n^{2(s-1)}, \qquad \phi_{2_i} := \varphi_i^2 \nu \nu_n^{2(s-1)},$$

where s,n>1. Multiplying the first equation of (3.1) by ϕ1iϕ_{1_i} and the second one by ϕ2iϕ_{2_i} respectively and integrating, we have

$$\int_{\Omega} |x-\xi_i|^{-2\gamma}\nabla u \nabla\phi_{1_i} = \frac{\sigma\alpha}{2^*}\int_{\Omega}|x-\xi_i|^{-2^*\gamma}u|u|^{\alpha-2}|\nu|^{\beta}\phi_{1_i} + \eta\int_{\Omega}|x-\xi_i|^{-2^*\gamma} \times u|u|^{2^*-2}\phi_{1_i} + a_1\int_{\Omega}|x-\xi_i|^{-2\gamma}u\phi_{1_i} + a_2\int_{\Omega}|x-\xi_i|^{-2\gamma} \times \nu\phi_{1_i} + \sum_{\substack{j=1\\ j\neq i}}^{k}\frac{\mu_j}{|x-\xi_j|^2}|x-\xi_i|^{-2\gamma}u\phi_{1_i}.$$

Note that $\nabla\phi_{1_i} = 2\varphi_i u u_n^{2(s-1)}\nabla\varphi_i + \varphi_i^2 u_n^{2(s-1)}\nabla u + 2(s-1)\varphi_i^2 u_n^{2(s-1)}\nabla u_n.$

Then

$$\int_\Omega |x-\xi_i|^{-2\gamma}\nabla u\nabla\phi_{1_i} = 2\int_\Omega |x-\xi_i|^{-2\gamma}\varphi_i u u_n^{2(s-1)}\nabla\varphi_i\nabla u + \int_\Omega |x-\xi_i|^{-2\gamma}\varphi_i^2 \times u_n^{2(s-1)}|\nabla u|^2 + 2(s-1)\int_\Omega |x-\xi_i|^{-2\gamma}\varphi_i^2 u_n^{2(s-1)}|\nabla u_n|^2.$$

By the Cauchy inequality and the Young inequality, we get

$$\left|2\int_\Omega |x-\xi_i|^{-2\gamma}\varphi_i u u_n^{2(s-1)}\nabla\varphi_i\nabla u\right| \le \frac{1}{2}\int_\Omega |x-\xi_i|^{-2\gamma}\varphi_i^2 u_n^{2(s-1)}|\nabla u|^2 + 2\int_\Omega |x-\xi_i|^{-2\gamma}|\nabla\varphi_i|^2 u^2 u_n^{2(s-1)}. \tag{3.2}$$

The same result holds for $\int_\Omega |x-\xi_i|^{-2\gamma}\nabla\phi_{2_i}\nabla v$.

By letting $\psi_1(x)=\varphi_i u u_n^{(s-1)}$, $\psi_2(x)=\varphi_i v v_n^{(s-1)}$, we have

$$\begin{aligned}
&\int_\Omega |x-\xi_i|^{-2\gamma}\left(|\nabla\psi_1|^2+|\nabla\psi_2|^2\right)\\
&\le C\Bigg(\int_\Omega |x-\xi_i|^{-2\gamma}|\nabla\varphi_i|^2 u^2 u_n^{2(s-1)} + \int_\Omega |x-\xi_i|^{-2\gamma}\nabla u\nabla\phi_{1_i}\\
&\quad + \int_\Omega |x-\xi_i|^{-2\gamma}|\nabla\varphi_i|^2 v^2 v_n^{2(s-1)} + \int_\Omega |x-\xi_i|^{-2\gamma}\nabla v\nabla\phi_{2_i}\Bigg)\\
&\le C\Bigg(\int_\Omega |x-\xi_i|^{-2\gamma}|\nabla\varphi_i|^2\left(u^2u_n^{2(s-1)}+v^2v_n^{2(s-1)}\right)\\
&\quad + \int_\Omega |x-\xi_i|^{-2^*\gamma}\varphi_i^2|u|^\alpha|v|^\beta\left(u_n^{2(s-1)}+v_n^{2(s-1)}\right)\\
&\quad + \int_\Omega |x-\xi_i|^{-2^*\gamma}\varphi_i^2|u|^{2^*}u_n^{2(s-1)} + \int_\Omega |x-\xi_i|^{-2^*\gamma}\varphi_i^2|v|^{2^*}v_n^{2(s-1)}\\
&\quad + \sum_{\substack{j=1\\ j\ne i}}^k \mu_j\int_\Omega \frac{|x-\xi_i|^{-2\gamma}}{|x-\xi_j|^2}\varphi_i^2\left(u^2u_n^{2(s-1)}+v^2v_n^{2(s-1)}\right)\\
&\quad + \int_\Omega |x-\xi_i|^{-2\gamma}\varphi_i^2\left(u^2+v^2\right)\left(u_n^{2(s-1)}+v_n^{2(s-1)}\right)\Bigg).
\end{aligned} \tag{3.3}$$

Using Caffarelli-Kohn-Nirenberg inequality [10], we infer that

$$\begin{aligned}
&\left(\int_\Omega \frac{|\varphi_i u u_n^{(s-1)}|^{2^*}}{|x-\xi_i|^{2^*\gamma}}\right)^{\frac{2}{2^*}} + \left(\int_\Omega \frac{|\varphi_i v v_n^{(s-1)}|^{2^*}}{|x-\xi_i|^{2^*\gamma}}\right)^{\frac{2}{2^*}} \\
&\leq C\left(\int_\Omega |x-\xi_i|^{-2\gamma}\left|\nabla\left(\varphi_i u u_n^{(s-1)}\right)\right|^2 + \int_\Omega |x-\xi_i|^{-2\gamma}\left|\nabla\left(\varphi_i v v_n^{(s-1)}\right)\right|^2\right) \\
&= C\int_\Omega |x-\xi_i|^{-2\gamma}\left(|\nabla\psi_1|^2 + |\nabla\psi_2|^2\right) \\
&\leq C\left(\int_\Omega |x-\xi_i|^{-2\gamma}|\nabla\varphi_i|^2\left(u^2 u_n^{2(s-1)} + v^2 v_n^{2(s-1)}\right)\right. \\
&\quad + \int_\Omega |x-\xi_i|^{-2^*\gamma}\varphi_i^2|u|^\alpha|v|^\beta\left(u_n^{2(s-1)} + v_n^{2(s-1)}\right) \\
&\quad + \int_\Omega |x-\xi_i|^{-2^*\gamma}\varphi_i^2|u|^{2^*}u_n^{2(s-1)} + \int_\Omega |x-\xi_i|^{-2^*\gamma}\varphi_i^2|v|^{2^*}v_n^{2(s-1)} \\
&\quad + \sum_{\substack{j=1\\ j\neq i}}^{k}\mu_j\int_\Omega \frac{|x-\xi_i|^{-2\gamma}}{|x-\xi_j|^2}\varphi_i^2\left(u^2u_n^{2(s-1)} + v^2v_n^{2(s-1)}\right) \\
&\quad \left. + \int_\Omega |x-\xi_i|^{-2\gamma}\varphi_i^2\left(u^2+v^2\right)\left(u_n^{2(s-1)} + v_n^{2(s-1)}\right)\right).
\end{aligned} \tag{3.4}$$

Define

$$\omega(x) := \max\{u(x), v(x)\}, \qquad \omega_n(x) := \min\{\omega(x), n\}.$$

Then ωn(x):=max{un(x),νn(x)}. Now, from the Hölder inequality, we deduce that

$$\int_\Omega \frac{\varphi_i^2|\omega|^{2^*}\omega_n^{2(s-1)}}{|x-\xi_i|^{2^*\gamma}} \leq \left(\int_\Omega \frac{|\varphi_i\omega\omega_n^{(s-1)}|^{2^*}}{|x-\xi_i|^{2^*\gamma}}\right)^{\frac{2}{2^*}}\left(\int_{B_R(\xi_i)} \frac{|\omega|^{2^*}}{|x-\xi_i|^{2^*\gamma}}\right)^{\frac{2}{N}}, \tag{3.5}$$

$$\int_\Omega \frac{\varphi_i^2|\omega|^2\omega_n^{2(s-1)}}{|x-\xi_i|^{2\gamma}} \leq \left(\int_\Omega \frac{|\varphi_i\omega\omega_n^{(s-1)}|^{2^*}}{|x-\xi_i|^{2^*\gamma}}\right)^{\frac{2}{2^*}}\left|B_R(\xi_i)\right|^{\frac{2}{N}}. \tag{3.6}$$

In the sequel, we have

$$\int_\Omega |x-\xi_i|^{-2^*\gamma}\varphi_i^2|u|^\alpha|v|^\beta\left(u_n^{2(s-1)}+v_n^{2(s-1)}\right)$$
$$\le\int_\Omega |x-\xi_i|^{-2^*\gamma}\varphi_i^2|\omega|^\alpha|\omega|^\beta\left(\omega_n^{2(s-1)}+\omega_n^{2(s-1)}\right)$$
$$=\int_\Omega |x-\xi_i|^{-2^*\gamma}\varphi_i^2|\omega|^{2^*}\omega_n^{2(s-1)}$$
$$\le 2\left(\int_\Omega \frac{|\varphi_i\omega\omega_n^{(s-1)}|^{2^*}}{|x-\xi_i|^{2^*\gamma}}\right)^{\frac{2}{2^*}}\left(\int_{B_R(\xi_i)}\frac{|\omega|^{2^*}}{|x-\xi_i|^{2^*\gamma}}\right)^{\frac{2}{N}}. \tag{3.7}$$

By the choice of φiφ_i, we obtain

$$\sum_{\substack{j=1\\ j\neq i}}^{k}\mu_j\int_\Omega \frac{|x-\xi_i|^{-2\gamma}}{|x-\xi_j|^2}\varphi_i^2\left(u^2u_n^{2(s-1)}+v^2v_n^{2(s-1)}\right)$$
$$\le C\int_\Omega |x-\xi_i|^{-2\gamma}\varphi_i^2\left(u^2u_n^{2(s-1)}+v^2v_n^{2(s-1)}\right)$$
$$\le\int_\Omega |x-\xi_i|^{-2\gamma}\varphi_i^2\omega^2\omega_n^{2(s-1)}$$
$$\le 2C\left(\int_\Omega \frac{|\varphi_i\omega\omega_n^{(s-1)}|^{2^*}}{|x-\xi_i|^{2^*\gamma}}\right)^{\frac{2}{2^*}}|B_R(\xi_i)|^{\frac{2}{N}}. \tag{3.8}$$

So, from (3.4) to (3.8) it follows that

$$\left(\int_\Omega |x-\xi_i|^{-2^*\gamma}|\varphi_i\omega\omega_n^{s-1}|^{2^*}\right)^{\frac{2}{2^*}}$$
$$\le C\int_\Omega |x-\xi_i|^{-2\gamma}|\nabla\varphi_i|^2\omega^2\omega_n^{2(s-1)}. \tag{3.9}$$

Take s=2∗2 and φi(x) to be a constant near the zero. Letting n→∞, we infer that ω∈ $L^{\frac{2^{*2}}{2}}(B_R(\xi_i),|x-\xi_i|^{-2^*\gamma})$ and so

$$u, v \in L^{\frac{2^{*2}}{2}}\left(B_R(\xi_i), |x-\xi_i|^{-2^*\gamma}\right). \tag{3.10}$$

Suppose r>0$r > 0$ is sufficiently small such that $r + l < R$ and φi is a cut-off function with the properties |∇φi|<1l and φi(x)=1 in Br(ξi).

Set t:=2∗22(2∗−2), δ:=2γt−2∗γ(t−1).

Then we have the following results:

$$\begin{aligned}
&\int_\Omega |x-\xi_i|^{-2\gamma}|\nabla\varphi_i|^2\omega^2\omega_n^{2(s-1)} \\
&\quad \le Cl^{-2}\int_\Omega |x-\xi_i|^{-(2^*\gamma\frac{t-1}{t})}\omega^2\omega_n^{2(s-1)}|x-\xi_i|^{-\gamma(2-2^*\frac{t-1}{t})} \\
&\quad \le Cl^{-2}\left(\int_{B_{r+l}(\xi_i)}\frac{|\omega\omega_n^{(s-1)}|^{\frac{2t}{t-1}}}{|x-\xi_i|^{2^*\gamma}}\right)^{\frac{t-1}{t}}\left(\int_{B_{r+l}(\xi_i)}\frac{1}{|x-\xi_i|^{\delta}}\right)^{\frac{1}{t}} \\
&\quad \le Cl^{-2}\left(\int_{B_{r+l}(\xi_i)}\frac{|\omega\omega_n^{(s-1)}|^{\frac{2t}{t-1}}}{|x-\xi_i|^{2^*\gamma}}\right)^{\frac{t-1}{t}},
\end{aligned} \tag{3.11}$$

where we used the Hölder inequality. From (3.9) in combination with (3.11), it follows that

$$\left(\int_{B_{r+l}(\xi_i)}|x-\xi_i|^{2^*\gamma}|\omega|^{2^*s}\right)^{\frac{1}{2^*s}} \le C^{\frac{1}{2s}}l^{-\frac{1}{2s}}\left(\int_{B_{r+l}(\xi_i)}|x-\xi_i|^{2^*\gamma}|\omega|^{\bar{p}_0 s}\right)^{\frac{1}{\bar{p}_0 s}}, \tag{3.12}$$

where $\bar{p}_0 = \frac{2t}{t-1} < 2^*$.

Denote s=χj, χ=pp¯0 and l=ρj, j≥1, where χ≥1, 2∗γ<N and p¯0χj=pχj−1. Using (3.12) recursively, we get

$$\begin{aligned}
\left(\int_{B_r(\xi_i)}|\omega|^{2^*\chi^j}\right)^{\frac{1}{2^*\chi^j}} &\le r^{\frac{\gamma}{\chi^j}}\left(\int_{B_r(\xi_i)}\frac{|\omega|^{2^*\chi^j}}{|x-\xi_i|^{2^*\gamma}}\right)^{\frac{1}{2^*\chi^j}} \\
&\le r^{\frac{\gamma}{\chi^j}}C^{\sum_{i=1}^{j}\frac{1}{2\chi^j}}\rho^{-\sum_{i=1}^{j}\frac{i}{\chi^i}}\left(\int_{B_{r+\rho}(\xi_i)}|x-\xi_i|^{2^*\gamma}|\omega|^{2^*}\right)^{\frac{1}{2^*}},
\end{aligned}$$

we have χj→∞ as j→∞. Note that the infinite sums on the right-hand side converge, then we obtain that ω∈L∞(Br(ξi)), particularly, we have u,v∈L∞(Ω). Thus,

$$u_0(x) = |x-\xi_i|^{-\gamma}u(x) \le M_1|x-\xi_i|^{-\gamma} \quad \text{for } x \in B_r(\xi_i)\setminus\{\xi_i\},$$

where $M_1 = \max\{\|u\|_{L^\infty(B_r(\xi_i))}, 1 \le i \le k\}$.

$$v_0(x) = |x-\xi_i|^{-\gamma}v(x) \le M_2|x-\xi_i|^{-\gamma} \quad \text{for } x \in B_r(\xi_i)\setminus\{\xi_i\},$$

where $M_1 = \max\{\|v\|_{L^\infty(B_r(\xi_i))}, 1 \le i \le k\}$. The proof is complete. □

Proof of Theorem 1.2 Suppose (u0,v0)∈H×H is a positive solution to problem (1.1). For all 0≤μi≤μ¯, set

$$u(x) = |x-\xi_i|^{\gamma}u_0(x) \quad \text{and} \quad v(x) = |x-\xi_i|^{\gamma}v_0(x).$$

Then

$$\begin{cases} -\operatorname{div}(|x-\xi_i|^{-2\gamma}\nabla u) \\ \quad = \frac{\sigma\alpha}{2^*}|x-\xi_i|^{-2^*\gamma}u^{\alpha-1}v^{\beta} + \eta|x-\xi_i|^{-2^*\gamma}u^{2^*-1} \\ \qquad + a_1|x-\xi_i|^{-2\gamma}u + a_2|x-\xi_i|^{-2\gamma}v + \sum_{\substack{j=1\\ j\ne i}}^{k}\frac{\mu_j}{|x-\xi_j|^2}|x-\xi_i|^{-2\gamma}u, \\ -\operatorname{div}(|x-\xi_i|^{-2\gamma}\nabla v) \\ \quad = \frac{\sigma\beta}{2^*}|x-\xi_i|^{-2^*\gamma}v^{\beta-1}u^{\alpha} + \lambda|x-\xi_i|^{-2^*\gamma}v^{2^*-1} \\ \qquad + a_2|x-\xi_i|^{-2\gamma}u + a_3|x-\xi_i|^{-2\gamma}v + \sum_{\substack{j=1\\ j\ne i}}^{k}\frac{\mu_j}{|x-\xi_j|^2}|x-\xi_i|^{-2\gamma}v. \end{cases} \tag{3.13}$$

Choose 0<ρ0<ρ and define n(t)=min|x−ξi|=tu(x) for ρ0≤t≤ρ. Let

$$n(\rho_0) = A\rho_0^{-2\sqrt{\bar\mu-\mu_i}} + B, \qquad n(\rho) = A\rho^{-2\sqrt{\bar\mu-\mu_i}} + B, \quad \text{where}$$

$$A = \frac{n(\rho)-n(\rho_0)}{\rho^{-2\sqrt{\bar\mu-\mu_i}} - \rho_0^{-2\sqrt{\bar\mu-\mu_i}}}, \qquad B = \frac{n(\rho_0)\rho^{-2\sqrt{\bar\mu-\mu_i}} - n(\rho)\rho_0^{-2\sqrt{\bar\mu-\mu_i}}}{\rho^{-2\sqrt{\bar\mu-\mu_i}} - \rho_0^{-2\sqrt{\bar\mu-\mu_i}}}.$$

It is easy to verify that

$$-\operatorname{div}\left(|x-\xi_i|^{-2(\sqrt{\bar{\mu}}-\sqrt{\bar{\mu}-\mu_i})}\nabla\left(A|x-\xi_i|^{-2\sqrt{\bar{\mu}-\mu_i}}+B\right)\right)=0 \quad \forall x\in\Omega\backslash\{\xi_i\}. \tag{3.14}$$

Combining (3.13) with (3.14), we get

$$-\operatorname{div}\left(|x-\xi_i|^{-2(\sqrt{\bar{\mu}}-\sqrt{\bar{\mu}-\mu_i})}\nabla\left(u-A|x-\xi_i|^{-2\sqrt{\bar{\mu}-\mu_i}}+B\right)\right)\geq 0, \quad \forall x\in B_\rho(\xi_i)\backslash B_{\rho_0}(\xi_i),$$
$$u(x)-\left(A|x-\xi_i|^{-2\sqrt{\bar{\mu}-\mu_i}}+B\right)\geq 0, \quad \forall x\in\partial\left(B_\rho(\xi_i)\backslash B_{\rho_0}(\xi_i)\right).$$

Therefore, by the maximum principle in Bρ(ξi)\Bρ0(ξi), we obtain

$$u(x)-\left(A|x-\xi_i|^{-2\sqrt{\bar{\mu}-\mu_i}}+B\right)\geq 0, \quad \forall x\in B_\rho(\xi_i)\backslash B_{\rho_0}(\xi_i).$$

Thus, for all x∈Bρ(ξi)\Bρ0(ξi),

$$\begin{aligned}u(x)&\geq A|x-\xi_i|^{-2\sqrt{\bar{\mu}-\mu_i}}+B\\&=\frac{|x-\xi_i|^{-2\sqrt{\bar{\mu}-\mu_i}}-\rho^{-2\sqrt{\bar{\mu}-\mu_i}}}{\rho_0^{-2\sqrt{\bar{\mu}-\mu_i}}-\rho^{-2\sqrt{\bar{\mu}-\mu_i}}}n(\rho_0)+\frac{\rho_0^{-2\sqrt{\bar{\mu}-\mu_i}}-|x-\xi_i|^{-2\sqrt{\bar{\mu}-\mu_i}}}{\rho_0^{-2\sqrt{\bar{\mu}-\mu_i}}-\rho^{-2\sqrt{\bar{\mu}-\mu_i}}}n(\rho)\\&\geq\frac{|x-\xi_i|^{2\sqrt{\bar{\mu}-\mu_i}}-\rho_0^{2\sqrt{\bar{\mu}-\mu_i}}}{|x-\xi_i|^{2\sqrt{\bar{\mu}-\mu_i}}-\rho_0^{2\sqrt{\bar{\mu}-\mu_i}}\rho^{-2\sqrt{\bar{\mu}-\mu_i}}|x-\xi_i|^{2\sqrt{\bar{\mu}-\mu_i}}}n(\rho).\end{aligned}$$

Taking ρ0→0, we conclude u(x)≥n(ρ)=min|x−ξi|=ρu(x)>0 for all x∈Bρ(ξi)\{ξi}.

Similar result also holds for v(x). Therefore, we have

$$\begin{aligned}u_0(x)&=|x-\xi_i|^{-\gamma}u(x)\geq|x-\xi_i|^{-\gamma}\min_{|x-\xi_i|=\rho}u(x)=|x-\xi_i|^{-\gamma}C_i\\&\geq|x-\xi_i|^{-\gamma}\min_{i=1,2,\ldots,k}C_i=|x-\xi_i|^{-\gamma}N_1.\end{aligned}$$

For any $x\in B_\rho(\xi_i)\backslash\{\xi_i\}$,

$$\nu_0(x) = |x-\xi_i|^{-\gamma}\nu(x) \geq |x-\xi_i|^{-\gamma} \min_{|x-\xi_i|=\rho} \nu(x) = |x-\xi_i|^{-\gamma}\acute{C}_i$$
$$\geq |x-\xi_i|^{-\gamma} \min_{i=1,2,\dots,k} \acute{C}_i = |x-\xi_i|^{-\gamma}N_2.$$

For any $x \in B_\rho(\xi_i)\backslash\{\xi_i\}$, . This proves the theorem.

4 LOCAL *(PS)C*-CONDITION AND THE EXISTENCE OF POSITIVE SOLUTIONS

We first establish a compactness result.

Lemma 4.1 *Suppose that* (H1) *holds. Then J satisfies the* (PS)c-*condition for all*

$$c < c^* := \frac{1}{N}\min\{S^{\frac{N}{2}}_{\eta,\lambda,\sigma}(\mu_1),\dots,S^{\frac{N}{2}}_{\eta,\lambda,\sigma}(\mu_k),(S_0)^{\frac{N}{2}}\} = \frac{1}{N}S^{\frac{N}{2}}_{\eta,\lambda,\sigma}(\mu_k).$$

Proof Suppose that {(un,vn)}⊂H×H satisfies J(un,vn)→c<c∗ and J′(un,vn)→0. The standard argument shows that {(un,vn)} is bounded in H×H$H \times H$.

For some (u,v)∈H×H, we have

$$(u_n,v_n) \rightharpoonup (u,v) \quad \text{weakly in } H \times H,$$
$$(u_n,v_n) \rightharpoonup (u,v) \quad \text{weakly in } L^2(\Omega,|x-\xi_i|^{-2}) \times L^2(\Omega,|x-\xi_i|^{-2}),$$
$$(u_n,v_n) \rightharpoonup (u,v) \quad \text{weakly in } L^{2^*}(\Omega) \times L^{2^*}(\Omega),$$
$$(u_n,v_n) \to (u,v) \quad \text{strongly in } L^{q_1}(\Omega) \times L^{q_2}(\Omega), q_1,q_2 \in [1,2^*),$$
$$(u_n,v_n) \to (u,v) \quad \text{a.e. in } \Omega.$$

Therefore, (u,v) is a solution to (1.1). Then by the concentration-compactness principle [11–13] and up to a subsequence, there exist an at most countable set J, a set of different points $\{x_j\}_{j\in\mathcal{J}} \subset \Omega\backslash\{\xi_i\}_{i=1}^k$, nonnegative real numbers $\tilde{\tau}_{x_j}, \tilde{\nu}_{x_j}, j \in \mathcal{J}$, and $\tilde{\tau}_{\xi_i}, \tilde{\nu}_{\xi_i}, \tilde{\gamma}_{\xi_i}$ $(1 \le i \le k)$ such that the following convergence holds in the sense of measures:

$$|\nabla u_n|^2 + |\nabla v_n|^2 \rightharpoonup d\tilde{\tau} \ge |\nabla u|^2 + |\nabla v|^2 + \sum_{j \in \mathcal{J}} \tilde{\tau}_{x_j} \delta_{x_j} + \sum_{i=1}^{k} \tilde{\tau}_{\xi_i} \delta_{\xi_i},$$

$$\frac{u_n^2 + v_n^2}{|x - \xi_i|^2} \rightharpoonup d\tilde{\gamma} = \frac{u^2 + v^2}{|x - \xi_i|^2} + \tilde{\gamma}_{\xi_i} \delta_{\xi_i},$$

$$\eta |u_n|^{2^*} + \lambda |v_n|^{2^*} + \sigma |u_n|^{\alpha} |v_n|^{\beta} \rightharpoonup d\tilde{\nu}$$

$$= \eta |u|^{2^*} + \lambda |v|^{2^*}$$

$$+ \sigma |u|^{\alpha} |v|^{\beta} + \sum_{j \in \mathcal{J}} \tilde{\nu}_{x_j} \delta_{x_j} + \sum_{i=1}^{k} \tilde{\nu}_{\xi_i} \delta_{\xi_i}.$$

By the Sobolev inequalities [10], we have

$$S_{\mu_i} \tilde{\nu}_{\xi_i}^{\frac{2}{2^*}} \le \tilde{\tau}_{\xi_i} - \mu_i \tilde{\gamma}_{\xi_i}, \quad 1 \le i \le k. \tag{4.1}$$

We claim that J is finite, and for any $j \in \mathcal{J}$, $\tilde{\nu}_{x_j} = 0$ or $\tilde{\nu}_{x_j} \ge S_0^{\frac{N}{2}}$.

In fact, let ε>0 be small enough for any $1 \le i \le k$, $\xi_i \notin B_\varepsilon(x_j)$ and $B_\varepsilon(x_i) \cap B_\varepsilon(x_j) = \emptyset$ for $i \neq j, i, j \in \mathcal{J}$. Let ϕ_ε^j be a smooth cut-off function centered at xj such that $0 \le \phi_\varepsilon^j \le 1$, $\phi_\varepsilon^j = 1$ $|x - x_j| \le \frac{\varepsilon}{2}$, $\phi_\varepsilon^j = 0$ for $|x - x_j| \ge \varepsilon$ and $|\nabla \phi_\varepsilon^j| \le \frac{4}{\varepsilon}$.

$$\lim_{\varepsilon\to 0}\lim_{n\to\infty}\int_\Omega \left(|\nabla u_n|^2 + |\nabla v_n|^2\right)\phi_\varepsilon^j = \lim_{\varepsilon\to 0}\int_\Omega \phi_\varepsilon^j\, d\tilde{\tau}$$

$$\geq \lim_{\varepsilon\to 0}\left(\int_\Omega \left(|\nabla u|^2 + |\nabla v|^2\right)\phi_\varepsilon^j + \tilde{\tau}_{x_j}\right)$$

$$= \tilde{\tau}_{x_j},$$

$$\lim_{\varepsilon\to 0}\lim_{n\to\infty}\int_\Omega \frac{u_n^2 + v_n^2}{|x-\xi_i|^2}\phi_\varepsilon^j = \lim_{\varepsilon\to 0}\int_\Omega \phi_\varepsilon^j\, d\tilde{\gamma} = \int_\Omega \frac{u^2+v^2}{|x-\xi_i|^2}\phi_\varepsilon^j = 0,$$

$$\lim_{\varepsilon\to 0}\lim_{n\to\infty}\int_\Omega \left(\eta|u_n|^{2^*} + \lambda|v_n|^{2^*} + \sigma|u_n|^\alpha|v_n|^\beta\right)\phi_\varepsilon^j$$

$$= \lim_{\varepsilon\to 0}\int_\Omega \phi_\varepsilon^j\, d\tilde{v}$$

$$= \lim_{\varepsilon\to 0}\left(\int_\Omega \left(\eta|u|^{2^*} + \lambda|v|^{2^*} + \sigma|u|^\alpha|v|^\beta\right)\phi_\varepsilon^j + \tilde{v}_{x_j}\right)$$

$$= \tilde{v}_{x_j},$$

$$\lim_{\varepsilon\to 0}\lim_{n\to\infty}\int_\Omega (u_n\nabla u_n + v_n\nabla v_n)\nabla\phi_\varepsilon^j = 0.$$

Then we have

$$0 = \lim_{\varepsilon\to 0}\lim_{n\to\infty}\left\langle J_\lambda'(u_n, v_n), (u_n\phi_\varepsilon^j, v_n\phi_\varepsilon^j)\right\rangle \geq \tilde{\tau}_{x_j} - \tilde{v}_{x_j}.$$

By the Sobolev inequality, $S_0\tilde{v}_{x_j}^{\frac{2}{2^*}} \leq \tilde{\tau}_{x_j}$;and then we deduce that $\tilde{v}_{x_j} = 0$ or $\tilde{v}_{x_j} \geq S_0^{\frac{N}{2}}$, , which implies that J is finite.

Now, we consider the possibility of concentration at points ξi (1≤i≤k), for ε>0 small enough that xj∉Bε(ξi) for all j∈J and Bε(ξi)∩Bε(ξj)=∅ for i≠j $i \neq j$ and 1≤i, j≤k. Let φiε be a smooth cut-off function centered at ξi such that 0≤φiε≤1, φiε=1 for |x−ξi|≥ε and |∇φiε|≤4ε. Then

$$\lim_{\varepsilon\to0}\lim_{n\to\infty}\int_\Omega (|\nabla u_n|^2+|\nabla v_n|^2)\phi_\varepsilon^j = \lim_{\varepsilon\to0}\int_\Omega \phi_\varepsilon^j\, d\tilde{\tau}$$
$$\geq \lim_{\varepsilon\to0}\left(\int_\Omega (|\nabla u|^2+|\nabla v|^2)\phi_\varepsilon^j + \tilde{\tau}_{x_j}\right)$$
$$= \tilde{\tau}_{x_j},$$

$$\lim_{\varepsilon\to0}\lim_{n\to\infty}\int_\Omega \frac{u_n^2+v_n^2}{|x-\xi_i|^2}\phi_\varepsilon^j = \lim_{\varepsilon\to0}\int_\Omega \phi_\varepsilon^j\, d\tilde{\gamma} = \int_\Omega \frac{u^2+v^2}{|x-\xi_i|^2}\phi_\varepsilon^j = 0,$$

$$\lim_{\varepsilon\to0}\lim_{n\to\infty}\int_\Omega (\eta|u_n|^{2^*}+\lambda|v_n|^{2^*}+\sigma|u_n|^\alpha|v_n|^\beta)\phi_\varepsilon^j$$
$$= \lim_{\varepsilon\to0}\int_\Omega \phi_\varepsilon^j\, d\tilde{\nu}$$
$$= \lim_{\varepsilon\to0}\left(\int_\Omega (\eta|u|^{2^*}+\lambda|v|^{2^*}+\sigma|u|^\alpha|v|^\beta)\phi_\varepsilon^j + \tilde{\nu}_{x_j}\right)$$
$$= \tilde{\nu}_{x_j},$$

$$\lim_{\varepsilon\to0}\lim_{n\to\infty}\int_\Omega (u_n\nabla u_n + v_n\nabla v_n)\nabla\phi_\varepsilon^j = 0.$$

Thus, we have

$$0 = \lim_{\varepsilon\to0}\lim_{n\to\infty}\langle J_\lambda'(u_n,v_n),(u_n\varphi_\varepsilon^i, v_n\varphi_\varepsilon^i)\rangle \geq \tilde{\tau}_{\xi_i} - \mu_i\tilde{\gamma}_{\xi_i} - \tilde{\nu}_{\xi_i}. \tag{4.2}$$

From (4.1) and (4.2) we derive that $S_{\mu_i}\tilde{\nu}_{\xi_i}^{\frac{2}{2^*}} \leq \tilde{\nu}_{\xi_i}, 1\leq i\leq k,$ and then either $\tilde{\nu}_{\xi_i}=0$ or $\tilde{\nu}_{\xi_i}\geq S_{\mu_i}^{\frac{N}{2}}$. On the other hand, from the above arguments, we conclude that

$$
\begin{aligned}
c &= \lim_{n\to\infty}\left(J(u_n,\nu_n)-\frac{1}{2}\langle J'(u_n,\nu_n),(u_n,\nu_n)\rangle\right)\\
&= \frac{1}{N}\lim_{n\to\infty}\int_\Omega\left(\eta|u_n|^{2^*}+\lambda|\nu_n|^{2^*}+\sigma|u_n|^{\alpha}|\nu_n|^{\beta}\right)dx\\
&= \frac{1}{N}\left(\int_\Omega\left(\eta|u|^{2^*}+\lambda|\nu|^{2^*}+\sigma|u|^{\alpha}|\nu|^{\beta}\right)dx+\sum_{j\in\mathcal{J}}\tilde{\nu}_{x_j}+\sum_{i=1}^{k}\tilde{\nu}_{\xi_i}\right)\\
&= \frac{1}{N}\left(\sum_{j\in\mathcal{J}}\tilde{\nu}_{x_j}+\sum_{i=1}^{k}\tilde{\nu}_{\xi_i}\right)+J(u,\nu).
\end{aligned}
$$

If $\tilde{\nu}_{\xi_i}=\tilde{\nu}_{x_j}=0$ for all i∈{1,...,k} and j∈J, then c=0$c=0$, which contradicts the assumption that c>0. On the other hand, if there exists an i∈{1,...,k}$i\in\{1,\dots,k\}$ such that v˜ξi≠0 or there exists a j∈J with v˜xj≠0, then we infer that

$$
c\geq\frac{1}{N}\min\left\{\left(S_{\eta,\lambda,\sigma}(0)\right)^{N/2},\left(S_{\eta,\lambda,\sigma}(\mu_1)\right)^{N/2},\dots,\left(S_{\eta,\lambda,\sigma}(\mu_k)\right)^{N/2}\right\}=\frac{1}{N}\left(S_{\eta,\lambda,\sigma}(\mu_k)\right)^{N/2},
$$

which contradicts our assumptions. Hence, (un,vn)→(u,v), as n→∞ in H×H. □

First, under the assumptions (H1), (H2), we have the following notations:

$$
f_{\eta,\lambda,\sigma}(\tau)=\frac{(1+\tau^2)S_{\mu_k}}{(\eta+\sigma\tau^{\beta}+\lambda\tau^{2^*})^{\frac{2}{2^*}}},\qquad \tau>0;
$$

$$
f_{\eta,\lambda,\sigma}(\tau_{\min}):=\min_{\tau>0}f_{\eta,\lambda,\sigma}(\tau)>0,\qquad \sigma>0,
$$

where τmin>0 is a minimal point of fη,λ,σ(τ), and therefore a root of the equation

$$
\alpha\sigma\tau^{\beta}-\sigma\beta\tau^{\beta-2}-2^*\lambda\tau^{2^*-2}+2^*\eta=0,\qquad \tau>0.
$$

Lemma 4.2 *Suppose that* (H1$\mathcal{H}_1$) *holds. Then we have* *(i)* $S_{\eta,\lambda,\sigma}(\mu) = f_{\eta,\lambda,\sigma}(\tau_{\min})$

(ii)Sη,λ,σ(μ) has the minimizers $(V^{\xi}_{\mu,\varepsilon}(x), \tau_{\min} V^{\xi}_{\mu,\varepsilon}(x)), \forall \varepsilon > 0$, *where* $V^{\xi}_{\mu,\varepsilon}(x)$ *are the extremal functions of* Sη,λ,σ(μ) *defined as in* (2.2).

Proof The argument is similar to that of [6].

Lemma 4.3 *Under the assumptions of* (H1), *we have*

$$\sup J\left(tu_{\varepsilon,\mu_k}, t(\tau_{\min} u_{\varepsilon,\mu_k})\right) < c^* = \frac{1}{N}\left(S_{\eta,\lambda,\sigma}(\mu_k)\right)^{N/2}.$$

Proof Suppose (H1) holds. Define the function

$$g(t) := J\left(tu_{\varepsilon,\mu_k}, t(\tau_{\min} u_{\varepsilon,\mu_k})\right), \quad t \geq 0.$$

Note that limt→+∞g(t)=−∞ and g(t)>0 as *t* is close to 0. Thus, supt≥0g(t) is attained at some finite tε>0 with g′(tε)=0. Furthermore, c′<tε<c″, where c′ and c″ are the positive constants independent of ε. By using (1.2), we have

$$g(t) \leq \frac{t^2}{2}(1+\tau_{\min}^2)\left(\int_\Omega \left(|\nabla u_{\varepsilon,\mu_k}|^2 - \mu_k \frac{u^2_{\varepsilon,\mu_k}}{|x-\xi_i|^2} - \lambda_1 u^2_{\varepsilon,\mu_k}\right) dx\right)$$
$$- \frac{t^{2^*}}{2^*}\left(\sigma\tau_{\min}^{\beta} + \eta + \lambda\tau_{\min}^{2^*}\right)\int_\Omega |u_{\varepsilon,\mu_k}|^{2^*} dx.$$

Note that

$$\max\left(\frac{t^2}{2}B_1 - \frac{t^{2^*}}{2^*}B_2\right) = \frac{1}{N}\left(B_1 B_2^{-2/2^*}\right)^{N/2}, \quad B_1 > 0, B_2 > 0, \tag{4.3}$$

and $0 \leq \mu \leq \bar{\mu} - 1$ and so $2 < 2\sqrt{\bar{\mu} - \mu}$.

From (4.3), Lemma 4.2 and Lemma 4.3, it follows that

$$
\begin{aligned}
g(t_\varepsilon) &\le \frac{1}{N}\left(\frac{(1+\tau_{\min}^2)\int_\Omega(|\nabla u_{\varepsilon,\mu_k}|^2-\mu_k\frac{u_{\varepsilon,\mu_k}^2}{|x-\xi_k|^2}-\lambda_1 u_{\varepsilon,\mu_k}^2)\,dx}{((\sigma\tau_{\min}^{\beta}+\eta+\lambda\tau_{\min}^{2^*})\int_\Omega|u_{\varepsilon,\mu_k}|^{2^*}dx)^{2/2^*}}\right)^{N/2}\\
&\le \frac{1}{N}\left(\frac{f(\tau_{\min})}{S(\mu_k)}\times\frac{(S(\mu_k))^{N/2}+O(\varepsilon^{2\sqrt{\bar\mu-\mu}})-C\varepsilon^2}{(S(\mu_k))^{(N-2)/2}+O(\varepsilon^{2\sqrt{\bar\mu-\mu}})}\right)\\
&\le \frac{1}{N}\big(f(\tau_{\min})\big)^{N/2}+O\big(\varepsilon^{2\sqrt{\bar\mu-\mu}}\big)-C\varepsilon^2\\
&< \frac{1}{N}\big(S_{\eta,\lambda,\sigma}(\mu_k)\big)^{N/2}=c^*,
\end{aligned}
$$

so $g(t\varepsilon)<c*$. Hence, $g(t\varepsilon)<c*$, $\forall t\ge 0$ and

$$
\sup_{t\ge 0} g(t)=\sup_{t\ge 0} J\big(tu_{\varepsilon,\mu_k}, t(\tau_{\min}u_{\varepsilon,\mu_k})\big)<c^*, \quad \text{if } \mu<\bar\mu-1. \tag{4.4}
$$

Proof of Theorem 1.3 Set $c := \inf_{h\in\Gamma}\max_{t\in[0,1]} J(h(t))$, where

$$
\Gamma=\{h\in C([0,1],H\times H)\,|\,h(0)=(0,0), J(h(1))<0\}.
$$

Suppose that (H1) holds. For all $(u,v)\in H\times H\backslash\{(0,0)\}$, from the Young and Hardy-Sobolev inequalities, it follows that

$$
\begin{aligned}
J(u,v) &\ge C(\|u\|^2+\|v\|^2)-C(\|u\|^{2^*}+\|v\|^{2^*})\\
&\ge C\|(u,v)\|^2-C\|(u,v)\|^{2^*},
\end{aligned}
$$

and there exists a constant $\rho>0$ small such that

$$
b:=\inf_{\|(u,v)\|=\rho} J(u,v)>0=J(0,0).
$$

Since $J(tu,tv)\to-\infty$ as $t\to\infty$, there exists $t0>0$ $t_0>0$ such that $\|(t0u,t0v)\|>\rho$ and $J(t0u,t0v)<0$. By the mountain-pass theorem [14], there exists a sequence $\{(un,vn)\}\subset H\times H$ such that $J(un,vn)\to c$ and $J'(un,vn)\to 0$, as $n\to\infty$.

From Lemma 4.2 it follows that

$$0 < c \leq \sup_{t\in[0,1]} J(tt_0 u_{\varepsilon,\mu_k}, tt_0 \tau_{\min} u_{\varepsilon,\mu_k})$$

$$\leq \sup_{t\geq 0} J(tu_{\varepsilon,\mu_k}, t\tau_{\min} u_{\varepsilon,\mu_k})$$

$$< c^*.$$

By Lemma 4.1 there exists a subsequence of {(un,vn)}, still denoted by {(un,vn)}, such that (un,vn)→(u,v) strongly in H×H$H \times H$. Thus, we get a critical point (u,v) of *J* satisfying (1.1), and *c* is a critical value. Set u+=max{u,0}.

Replacing respectively *u*, *v* with u+ and v+ in terms of the right-hand side of (1.1) and repeating the above process, we can get a nonnegative nontrivial solution (u,v) of (1.1). If u≡0, we get v≡0 by (1.1) and the assumption a2>0. Similarly, if v≡0, we also have u≡0. There, u,v≢0. From the maximum principle, it follows that u,v>0 in Ω.

COMPETING INTERESTS

The authors declare that they have no competing interests.

AUTHORS' CONTRIBUTIONS

Each of the authors, SK, MF and OKK contributed to each part of this work equally and read and approved the final version of the manuscript.

REFERENCES

1. Cao D, Han P: Solutions to critical elliptic equations with multi-singular inverse square potentials. *J. Differ. Equ.* 2006, 224: 332-372.

2. Hsu TS: Multiple positive solutions for semilinear elliptic equations involving multi-singular inverse square potentials and concave-convex nonlinearities. *Nonlinear Anal.* 2011, 74: 3703-3715.

3. Kang D: On the weighted elliptic problems involving multi-singular potentials and multi-critical exponents. *Acta Math. Sin. Engl. Ser.* 2009, 25: 435-444.

4. Abdellaoui B, Felli V, Peral I:Some remarks on systems of elliptic equations doubly critical in the whole RN$\mathbb{R}^N$. *Calc. Var. Partial Differ. Equ.*2009, 34: 97-137.

5. Bouchekif M, Nasri Y: On a singular elliptic system at resonance. *Ann. Mat. Pura Appl.* 2010, 189: 227-240.

6. Huang Y, Kang D: On the singular elliptic systems involving multiple critical Sobolev exponents. *Nonlinear Anal.* 2011, 74: 400-412.

7. Kang D: Semilinear systems involving multiple critical Hardy-Sobolev exponents and three singular points. *Appl. Math.* 2011, 218: 4514-4522.

8. Kang D, Peng S: Existence and asymptotic properties of solutions to elliptic systems involving multiple critical exponents. *Sci. China Math.*2011, 54(2):243-256.

9. Kang D, Huang Y, Liu S: Asymptotic estimates on the extremal functions of a quasi-linear elliptic problem. *J. South-Central Univ. Natl. Nat. Sci. Ed.* 2008, 27(3):91-95.

10. Caffarelli L, Kohn R, Nirenberg L: First order interpolation inequality with weights. *Compos. Math.* 1984, 53: 259-275.

11. Cai, M, Kang, D: Concentration-compactness principles for the systems of critical elliptic equations. Acta Math. Sci. Ser. B Engl. Ed. (to appear)

12. Lions PL: The concentration-compactness principle in the calculus of variations: the limit case I. *Rev. Mat. Iberoam.* 1985, 1: 45-121.

13. Lions PL: The concentration-compactness principle in the calculus of variations: the limit case II. *Rev. Mat. Iberoam.* 1985, 1: 145-201.

14. Ambrosetti A, Rabinowitz PH: Dual variational methods in critical point theory and applications. *J. Funct. Anal.* 1973, 14: 349-381.

11

MULTIPLE SOLUTIONS OF SEMILINEAR ELLIPTIC SYSTEMS ON THE HEISENBERG GROUP

ABSTRACT

In this paper, a class of semilinear elliptic systems which have a strong resonance at the first eigenvalue on the Heisenberg group is considered. Under certain assumptions, by virtue of the variational methods, the multiple weak solutions of the systems are obtained.

KEYWORDS

semilinear elliptic system, strong resonance, variational method, Heisenberg group

1 INTRODUCTION

Let H^N be the space $R^N \times R^N \times R$ equipped with the following group operation:

$$\eta \circ \eta' = (x,y,t) \circ (x',y',t') = (x+x', y+y', t+t'+2(x' \cdot y - x \cdot y')),$$

where '·' denotes the usual inner-product in R^N. This operation endows H^N with the structure of a Lie group. The vector fields X1,...,XN, Y1,...,YN, *T*, given by

$$X_j = \frac{\partial}{\partial x_j} + 2y_j\frac{\partial}{\partial t}, \qquad Y_j = \frac{\partial}{\partial y_j} - 2x_j\frac{\partial}{\partial t}, \qquad T = \frac{\partial}{\partial t},$$

form a basis for the tangent space at η=(x,y,t).

Definition 1.1 The Heisenberg Laplacian is by definition

$$\Delta_H = \sum_{j=1}^{N}(X_j^2 + Y_j^2),$$

and let ∇Hu denote the 2*N*-vector (X1u,...,XNu,Y1u,...,YNu).

Definition 1.2 The space $S_0^{1,2}(\Omega)$ is defined as the completion of $C_0^{\infty}(\Omega)$ in the norm

$$\|u\|^2_{S_0^{1,2}} = \int_\Omega \sum_{j=1}^{N}(|X_ju|^2 + |Y_ju|^2) = \int_\Omega |\nabla_H u|^2.$$

Some existence and nonexistence for the semilinear equations or systems on the Heisenberg group have been studied by Garofalo, Lanconelli and Niu, see [1, 2], *etc.*

In this paper, we study the problems on the existence and multiplicity of solutions for the system

$$\begin{cases} -\Delta_H\binom{u}{v} = \lambda_1\begin{pmatrix} a(x) & b(x) \\ b(x) & d(x)\end{pmatrix}\binom{u}{v} - \binom{f(x,u,v)}{g(x,u,v)}, & x \in \Omega, \\ u = v = 0, \quad x \in \partial\Omega, \end{cases} \tag{1.1}$$

where Ω⊆H^N is a bounded smooth domain, $a, b, d \in C^0(\overline{\Omega}, \mathbb{R})$ and $f, g \in C^1(\overline{\Omega} \times \mathbb{R}^2, \mathbb{R})$ and $F(x,u,v) \in C^2(\overline{\Omega} \times \mathbb{R}^2, \mathbb{R})$. Moreover, we assume that there is some function

$F(x,u,v) \in C^2(\overline{\Omega} \times \mathbb{R}^2, \mathbb{R})$ such that ∇F=(fg). Here ∇F denotes the gradient in the variable $u$and v, $i.e., \frac{\partial F}{\partial u} = f, \frac{\partial F}{\partial v} = g.$.

In fact, the condition in $\mathbb{R}^N$ was studied by da Silva; we can see [3]. In this paper we study the problem on the Heisenberg group H^N. The elliptic problems at resonance have been studied by many authors; see [4–7].

We use the variation methods to solve problem (1.1). Finding weak solutions of (1.1) in $E = S_0^{1,2}(\Omega) \times S_0^{1,2}(\Omega)$ is equivalent to finding critical points of the C2 functional given by

$$I(h) = \frac{1}{2}\|h\|^2 - \frac{1}{2}\int_\Omega \langle Ah, h\rangle + \int_\Omega F(x,h), \tag{1.2}$$

where

$$h \in E, \quad h = \begin{pmatrix} h^{(1)} \\ h^{(2)} \end{pmatrix}, \qquad \|h\|^2 = \int_\Omega \left|\nabla_H h^{(1)}\right|^2 + \left|\nabla_H h^{(2)}\right|^2,$$

and ⟨·,·⟩ denotes the usual inner product in R^2.

We introduce the eigenvalue problem with weights. Let us denote by A$\mathcal{A}$ the set of all continuous, cooperative and symmetric matrices A of order 2, given by

$$A(x) = \begin{pmatrix} a(x) & b(x) \\ b(x) & d(x) \end{pmatrix},$$

where the functions $a, b, d \in C(\overline{\Omega}, \mathbb{R})$ satisfy the following conditions:

(A_1) A(x) is cooperative, that is, b(x)≥0.

(A_2) There is an x0∈Ω such that a(x0)>0 or d(x0)>0.

Given A∈A(Ω), consider the weighted eigenvalue problem

$$\begin{cases} -\Delta_{\mathrm{H}}\binom{h^{(1)}}{h^{(2)}} = \lambda A(x)\binom{h^{(1)}}{h^{(2)}}, & \text{in } \Omega, \\ h^{(1)} = h^{(2)} = 0, \quad \text{on } \partial\Omega, \end{cases}$$

if $A \in A(\Omega)$. By virtue of the spectral theory for compact operators, we obtain the sequence of eigenvalues

$$0 < \lambda_1 < \lambda_2 \leq \lambda_3 \leq \cdots$$

such that $\lambda k \to +\infty$ as $k \to \infty$; see [6, 8, 9]. Here, each eigenvalue λk, $k \geq 1$ has finite multiplicity, and we have

$$\frac{1}{\lambda_k} = \sup\left\{ \int_\Omega \langle Ah, h \rangle, \|h\| = 1, h \in V_{k-1}^\perp \right\},$$

where $Vk = \mathrm{span}\{\Phi 1, \ldots, \Phi k\}$ with $k \geq 1$.

Remark 1.1

(1) $E = V_k \oplus V_k^\perp$ for $k \geq 1$.

(2) The following variational inequalities hold:

$$\|h\|^2 \leq \lambda_k \int_\Omega \langle Ah, h \rangle, \quad \forall h \in V_k, k \geq 1, \tag{1.3}$$

$$\|h\|^2 \geq \lambda_{k+1} \int_\Omega \langle Ah, h \rangle, \quad \forall h \in V_k^\perp, k \geq 0. \tag{1.4}$$

The variational inequalities will be used in the next section. We would like to mention that the $\Phi 1$ is positive in Ω. In the paper, without loss of generality, we assume that $\lambda 1 = 1$.

We now state the assumptions and the main results in this paper. Firstly, we define the following functions:

$$\begin{cases} T^+ = \liminf_{(u,v)\to(\infty,\infty)} F(x,u,v), & S^+ = \limsup_{(u,v)\to(\infty,\infty)} F(x,u,v), \\ T^- = \liminf_{(u,v)\to(-\infty,-\infty)} F(x,u,v), & S^- = \limsup_{(u,v)\to(-\infty,-\infty)} F(x,u,v). \end{cases} \tag{1.5}$$

The above functions belong to L1(Ω) and the limits are taken a.e. and uniformly in x∈Ω.

Now we make the following basic hypotheses:

(E_0) There exists $k \in C(\overline{\Omega})$such that

$$\lim_{|h|\to\infty} \nabla F(x,h) = 0, \qquad |F(x,h)| \le k(x), \quad \text{a.e. } x \in \Omega, \forall h \in \mathbb{R}^2.$$

(E_1) $F(x,h) \ge \frac{1}{2}(1-\lambda_2)\langle Ah,h\rangle + b_1|\Omega|^{-1}$, $b_1 \ge 0$, $\forall (x,h) \in \Omega \times \mathbb{R}^2$.
(E_2) $\langle Ah,h\rangle \ge 0$, $\forall (x,h) \in \Omega \times \mathbb{R}^2$.
(E_3) There exist $\alpha \in (0,1)$ and $\delta > 0$ such that

$$F(x,h) \ge \frac{1-\alpha}{2}\langle Ah,h\rangle, \quad \forall x \in \Omega \text{ and } |z| < \delta.$$

(E_4) $\int_\Omega S^+ \le 0$ and $\int_\Omega S^- \le 0$.
(E_5) There exists $t_0 \in \mathbb{R}$ such that

$$\int_\Omega F(x, t_0\Phi_1) < \min\left\{\int_\Omega T^+, \int_\Omega T^-\right\}.$$

(E_6) There are $t_1^- < 0$ and $t_1^+ > 0$ such that

$$\int_\Omega F(x, t_1^\pm \Phi_1) < \min\left\{\int_\Omega T^+, \int_\Omega T^-\right\}.$$

We can prove that the associated functional *J* has the saddle geometry. Actually, we have the following results.

Theorem 1.1 *Let* $\Omega \subseteq \mathbb{H}^N$ *be a bounded smooth domain,* $a(x), b(x), d(x) \in C^0(\overline{\Omega}, \mathbb{R})$ *and* $f(x,u,v), g(x,u,v) \in C^1(\overline{\Omega} \times \mathbb{R}^2, \mathbb{R})$. *Assume that there is some function* $F(x,u,v) \in C^2(\overline{\Omega} \times \mathbb{R}^2, \mathbb{R})$ *such that* $\frac{\partial F}{\partial u} = f, \frac{\partial F}{\partial v} = g$. *such that* $\frac{\partial F}{\partial u} = f, \frac{\partial F}{\partial v} = g$. *Furthermore, if the conditions* (E_0), (E_1), (E_2) *are satisfied, problem* (1.1) *has at least one solution* z1∈E$z_1 \in E$.

Remark 1.2 For the hypotheses ∇F(x,0,0)≡0 and F(x,0,0)≡0, problem (1.1) admits the trivial solution (u,v)=0. In this case, the main point is to assure the existence of nontrivial solutions.

Theorem 1.2 *Let* Ω⊆HN *be a bounded smooth domain,* a(x),b(x),d(x)∈C0(Ω¯,R) *and* f(x,u,v),g(x,u,v)∈C1(Ω¯×R2,R). *Assume that there is some function* F(x,u,v)∈C2(Ω¯× $\mathbb{R}^2, \mathbb{R})$ *such that* $\frac{\partial F}{\partial u} = f, \frac{\partial F}{\partial v} = g$. *Furthermore, if the conditions* (E_0), (E_2), (E_3), (E_4) *and* (E_5) *are satisfied, then problem* (1.1) *has at least two nontrivial solutions.*

Theorem 1.3 *Let* Ω⊆H^N *be a bounded smooth domain,* a(x),b(x),d(x)∈C0(Ω,R) *and* f(x,u,v),g(x,u,v)∈C1(Ω¯×R2,R). *Assume that there is some function* F(x,u,v)∈C2(Ω¯× $\mathbb{R}^2, \mathbb{R})$ *such that* $\frac{\partial F}{\partial u} = f, \frac{\partial F}{\partial v} = g$. *Furthermore, if the conditions* (E_0), (E_1), (E_2), (E_3), (E_4) *and* (E_6) *are satisfied, then problem* (1.1) *has at least three nontrivial solutions.*

2 PRELIMINARIES AND FUNDAMENTAL LEMMAS

In this section, we prove some lemmas needed in the proof of our main theorems.

We first introduce the Folland-Stein embedding theorem (see [10]) as follows.

Lemma 2.1 *Let* Ω⊆HN *be a bounded domain and let* $Q = 2N + 2$. *Then* $S_0^{1,2}(\Omega)$ *compactly embedding in* $L^p(\Omega)$, *where* $2 \le p < \frac{2Q}{Q-2}$.

To establish Lemmas 2.7 and 2.8, we introduce the following corollary of the Ekeland variation principle (see [11]).

Lemma 2.2 *X is a metric space,* I∈C1(X,R) *is bounded from below, which satisfies the* (PS)c *condition, then* c=infx∈XE(x) *is a critical value of E.*

Next, we describe some results under the geometry for the functional *I*.

Lemma 2.3 *Under hypotheses* (E_0) *and* (E_1), *the functional I has the following saddle geometry*:

(L3-1) I(h)→∞ *if* ‖h‖→∞ *with* $h \in V_1^{\perp}$.

(L3-2) *There is* α∈R *such that* I(h)≤α, ∀z∈V1.

(L3-3) I(h)≥b1, ∀z∈V⊥1. $h \in V_1^{\perp}$.

Proof (L3-1). From (1.2), (1.4) we have

$$I(h) \geq \frac{1}{2}\left(1 - \frac{1}{\lambda_2}\right)\|h\|^2 + \int_{\Omega} F(x,h), \quad h \in V_1^{\perp}.$$

Using (E_0), we have J(h)→∞, as ‖h‖→∞.

(L3-2). By simple calculation, we get

$$I(h) = \int_{\Omega} F(x,h), \quad h \in V_1.$$

By using (E_0), we have

$$I(h) = \int_{\Omega} F(x,h) \leq \int_{\Omega} k(x).$$

So, we choose $\alpha = \int_{\Omega} k(x)$.

(L3-3). By (E_1) and the variational inequality (1.4), we have

$$\begin{aligned} I(h) &= \frac{1}{2}\|h\|^2 - \frac{1}{2}\int_\Omega \langle Ah, h\rangle + \int_\Omega F(x,h) \\ &\geq \frac{1}{2}\|h\|^2 - \frac{\lambda_2}{2}\int_\Omega \langle Ah, h\rangle + b_1 \\ &\geq b_1, \quad \forall z \in V_1^\perp, \end{aligned}$$

the proof of this lemma is completed. □

Next, we prove the Palais-Smale conditions at some levels for the functional *I*. We recall that *I*: E→R is said to satisfy the Palais-Smale conditions at the level c∈R ((PS)c in short) if any sequence $\{h_n\} \subseteq E$ such that

$$I(h_n) \to c, \qquad I'(h_n) \to 0,$$

as n→∞$n \to \infty$, possesses a convergent subsequence in *E*. Moreover, we say that *I* satisfies the (PS) conditions when we have (PS)c for all c∈R.

Lemma 2.4 *Assume that the condition* (E_0) *holds. Then the functional I has the* (PS)c *conditions whenever* c<min $\min\{\int_\Omega T^+, \int_\Omega T^-\}$ *or* $c > \max\{\int_\Omega S^+, \int_\Omega S^-\}$.
or $c < \min\{\int_\Omega T^+, \int_\Omega T^-\}$.

Proof We only prove the condition for all $c < \min\{\int_\Omega T^+, \int_\Omega T^-\}$. For the case $c > \max\{\int_\Omega S^+, \int_\Omega S^-\}$, we can use similar methods.

*1.*Boundedness of the (PS) sequence.

The proof is by contradiction. Suppose that there exists a (PS)c unbounded sequence {hn}∈E such that $c < \min\{\int_\Omega T^+, \int_\Omega T^-\}$. For the ease of notation and without loss of generality, we assume that

$\|h_n\| \to \infty,$
$I(h_n) \to c,$
$I'(h_n) \to 0, \ n \to \infty.$

We define $\overline{h}_n = \frac{h_n}{\|h_n\|}$, hence there is an h¯∈E with the following properties:

$\overline{h}_n \rightharpoonup \overline{h}$ in E,
$\overline{h}_n \to \overline{h}$ in $L^p(\Omega) \times L^p(\Omega)$, where $2 \leq p < 2^*$ and $2^* = \frac{2N+2}{N}$,
$\overline{h}_n \to \overline{h}$ a.e. in Ω.

For any Φ∈E, obviously $\frac{I'(h_n)\Phi}{\|h_n\|} \to 0.$ By simple calculation, it is easy to obtain

$$I'(h_n)\Phi = \int_\Omega \langle \nabla_{\mathrm{H}} h_n, \nabla_H \Phi \rangle - \int_\Omega \langle A h_n, \Phi \rangle + \int_\Omega \langle \nabla F(x, h_n), \Phi \rangle,$$

where $h_n = \begin{pmatrix} h_n^{(1)} \\ h_n^{(2)} \end{pmatrix}, \ \Phi = \begin{pmatrix} \Phi^{(1)} \\ \Phi^{(2)} \end{pmatrix}.$ We have

$$\frac{I'(h_n)\Phi}{\|h_n\|} = \int_\Omega \langle \nabla_{\mathrm{H}} \overline{h}_n, \nabla_{\mathrm{H}} \Phi \rangle - \int_\Omega \langle A \overline{h}_n, \Phi \rangle + \frac{\int_\Omega \langle \nabla F(x, h_n), \Phi \rangle}{\|h_n\|} \to 0 \quad \text{as } n \to \infty.$$

From the convergence of $\{\overline{h}_n\}$, we have

$$\int_\Omega \langle \nabla_{\mathrm{H}} \overline{h}, \nabla_{\mathrm{H}} \Phi \rangle = \int_\Omega \langle A \overline{h}, \Phi \rangle.$$

We see that λ1=1, and by the definition of λ1, we obtain that h¯=±Φ1. So, we suppose initially that h¯=Φ1. Because Φ1 is positive, *i.e.*, Φ(1)1>0, Φ(2)1>0, it is obvious that h(1)n→∞, h(2)n→∞, ∀x∈Ω as n→∞.

Hence, we can take hn=tnΦ1+ωn, where {tn}∈R, {ωn}∈V⊥1, and we have

$$I(h_n) = \frac{1}{2}\|t_n\Phi_1 + \omega_n\|^2 - \frac{1}{2}\int_\Omega \langle A(t_n\Phi_1 + \omega_n), t_n\Phi_1 + \omega_n\rangle + \int_\Omega F(x, h_n)$$
$$= \frac{1}{2}\|\omega_n\|^2 - \int_\Omega \langle A\omega_n, \omega_n\rangle + \int_\Omega F(x, h_n).$$

Using (1.4), we obtain

$$I(h_n) \geq \frac{1}{2}\left(1 - \frac{1}{\lambda_2}\right)\|\omega_n\|^2 + \int_\Omega F(x, h_n). \tag{2.1}$$

Since I(hn)→c, it is easy to obtain that the sequence {ωn} is bounded. On the other hand, because of ||hn||→∞, on a subsequence |tn|→∞, without loss of generality, we assume tn→−∞.

Now, using Hölder's inequality and (E_0), we have

$$\left|\int_\Omega \nabla F(x, h_n)\omega_n\right| \leq C\left(\int_\Omega |\nabla F(x, h_n)|^2\right)^{\frac{1}{2}}.$$

Thus, applying the dominated convergence theorem, we conclude that

$$\lim_{n\to\infty}\int_\Omega \nabla F(x, h_n)\omega_n = 0. \tag{2.2}$$

On the other hand,

$$I'(h_n)\omega_n = \int_\Omega \langle \nabla_H h_n, \nabla_H \omega_n\rangle - \int_\Omega \langle Ah_n, \omega_n\rangle + \int_\Omega \langle \nabla F(x, h_n), \omega_n\rangle$$
$$= \|\omega_n\|^2 - \int_\Omega \langle Ah_n, \omega_n\rangle + \int_\Omega \langle \nabla F(x, h_n), \omega_n\rangle.$$

Using (2.2), (1.4), we obtain

$$\left(1 - \frac{1}{\lambda_2}\right)\|\omega_n\|^2 \leq |I'(h_n)\omega_n| + \left|\int_\Omega \langle \nabla F(x, h_n), \omega_n\rangle\right| \to 0$$

as n→∞. Therefore, by variational inequalities (1.3) and (1.4), we obtain that

$$\|\omega_n\|^2 - \int_\Omega \langle Ah_n, \omega_n\rangle \to 0 \quad \text{as } n \to \infty.$$

Consequently, by virtue of Fatou's lemma and (E_0), we have

$$c = \lim_{n\to\infty}\left(\frac{1}{2}\|\omega_n\|^2 - \int_\Omega \langle A\omega_n, \omega_n\rangle + \int_\Omega F(x, h_n)\right) = \liminf_{n\to\infty}\int_\Omega F(x, t_n\Phi_1 + \omega_n)$$
$$\geq \int_\Omega \liminf_{n\to\infty} F(x, t_n\Phi_1 + \omega_n) = \int_\Omega T^-,$$

which contradicts the condition $c < \min\{\int_\Omega T^+, \int_\Omega T^-\}$. Hence, the (PS)c sequence is bounded.

*2.*Various convergence of {hn}.

Since {hn} is a bounded sequence, there is an h∈E with the following properties:

$h_n \rightharpoonup h$ in E,
$h_n \to h$ in $L^p(\Omega) \times L^p(\Omega)$, where $2 \leq p < 2^*$ and $2^* = \frac{2N+2}{N}$,
$h_n \to h$ a.e. in Ω.

3.{hn} convergence to *h* in *E*.

From the definition of (PS)c sequence, we have, as n→∞,

$$I'(h_n)h = \int_\Omega \langle \nabla_H h_n, \nabla_H h\rangle - \int_\Omega \langle Ah_n, h\rangle + \int_\Omega \langle \nabla F(x, h_n), h\rangle \to 0,$$

$$I'(h_n)h_n = \int_\Omega |\nabla_H h_n|^2 - \int_\Omega \langle Ah_n, h_n\rangle + \int_\Omega \langle \nabla F(x, h_n), h_n\rangle \to 0.$$

By Fatou's lemma and the above convergence of {hn}, it is easy to show that

$$\int_\Omega \langle Ah_n, h\rangle \to \int_\Omega \langle Ah, h\rangle,$$

$$\int_\Omega \langle \nabla F(x,h_n), h\rangle \to \int_\Omega \langle \nabla F(x,h), h\rangle,$$

$$\int_\Omega \langle Ah_n, h_n\rangle \to \int_\Omega \langle Ah, h\rangle,$$

$$\int_\Omega \langle \nabla F(x,h_n), h_n\rangle \to \int_\Omega \langle \nabla F(x,h), h\rangle$$

as n→∞. Hence, we have

$$\int_\Omega \langle \nabla_H h_n, \nabla_H h\rangle \to \int_\Omega \langle Ah, h\rangle - \int_\Omega \langle \nabla F(x,h), h\rangle \quad \text{as } n \to \infty, \tag{2.3}$$

$$\int_\Omega |\nabla_H h_n|^2 \to \int_\Omega \langle Ah, h\rangle - \int_\Omega \langle \nabla F(x,h), h\rangle \quad \text{as } n \to \infty. \tag{2.4}$$

By weak convergence, we have

$$\int_\Omega \langle \nabla_H h_n, \nabla_H h\rangle \to \int_\Omega \langle \nabla_H h, \nabla_H h\rangle \quad \text{as } n \to \infty. \tag{2.5}$$

Using (2.3), (2.4) and (2.5), by simple calculation, we obtain

$$\int_\Omega |\nabla_H h_n - \nabla_H h|^2 \to 0 \quad \text{as } n \to \infty.$$

The proof is completed.

Lemma 2.5 *Suppose that* (E_0) *and* (E_3) *are satisfied. Then the origin is a local minimum for the functional I.*

Proof Using (E_3), we can choose $p \in (2, 2*)$ and a constant $C>0$ such that

$$F(x,h) \geq \frac{1-\alpha}{2}\langle Ah,h\rangle - C|h|^p, \quad \forall (x,h) \in \Omega \times \mathbb{R}^2.$$

Consequently, we have

$$\begin{aligned} I(h) &= \frac{1}{2}\|h\|^2 - \frac{1}{2}\int_\Omega \langle Ah,h\rangle + \int_\Omega F(x,h) \geq \frac{1}{2}(1-\alpha)\|h\|^2 - C\int_\Omega |h|^p \\ &\geq \frac{1}{2}(1-\alpha)\|h\|^2 - C\|h\|^p \geq \frac{1}{4}(1-\alpha)\|h\|^2, \quad \|h\| < \rho, \end{aligned}$$

where ρ is small enough and $0<\rho<t_0$, t_0 is provided by (E_5). Therefore the proof has been completed. □

To complete the mountain pass geometry, we prove the following result.

Lemma 2.6 *Let the hypotheses* (E_0), (E_4) *and* (E_5) *hold. Then there exist* $h_0 \in E$ *and* $\rho>0$ *such that* $I(h_0)<0$ *and* $\|h_0\|>\rho$.

Proof Using (E_2) and (E_5), we take $h_0=t_0\Phi_1$, where t_0 t_0 is provided by (E_5). Thus, we obtain

$$\begin{aligned} I(t_0\Phi_1) &= \frac{1}{2}\|t_0\Phi_1\|^2 - \frac{1}{2}\int_\Omega \left\langle A(t_0\Phi_1), t_0\Phi_1\right\rangle + \int_\Omega F(x, t_0\Phi_1), \\ &= \int_\Omega F(x,t_0\Phi_1) < \min\left\{\int_\Omega T^+, \int_\Omega T^-\right\} < \max\left\{\int_\Omega S^+, \int_\Omega S^-\right\} \leq 0, \end{aligned}$$

and $\|t_0\Phi_1\|=t_0$. If we take $0<\rho<t_0$, then the conclusion follows. □

Lemma 2.7 *Under hypotheses* (E_0), (E_4) *and* (E_5), *problem* (1.1) *has at least one nontrivial solution* $h_0 \in E$. *Moreover,* h_0 h_0 *has negative energy, i.e.,* $J(h_0)<0$.

Proof By (E_0) and (1.4), we obtain

$$I(h) = \frac{1}{2}\|h\|^2 - \frac{1}{2}\int_\Omega \langle Ah,h\rangle + \int_\Omega F(x,h) \geq \int_\Omega F(x,h) \geq -\int_\Omega k(x).$$

Therefore, the functional I is bounded below. In this case, we would like to mention that I has the (PS)c conditions with c=inf{I(h):h∈E}. For seeing this, by Lemma 2.4, we take t0∈R provided by (E_5) we can obtain

$$c \leq I(t\Phi_1) = \int_\Omega F(x, t\Phi_1) < \min\left\{\int_\Omega T^+, \int_\Omega T^-\right\} \leq 0.$$

Consequently, applying Lemma 2.2, we have one critical point h0∈E$h_0 \in E$ such that I(h0)=inf{I(h):h∈E}≤I(tΦ1)<0. The proof of this lemma is completed. □

To prove Theorem 1.3, we establish the following lemma.

Lemma 2.8 *Assume that the conditions* (E_0), (E_1), (E_4) *and* (E_6) *hold. Then problem* (1.1) *has at least two nontrivial solutions with negative energy.*

Proof Define

$$M^+ = \left\{t\Phi_1 + \omega, t \geq 0, \omega \in V_1^\perp\right\}, \qquad M^- = \left\{t\Phi_1 + \omega, t \leq 0, \omega \in V_1^\perp\right\}.$$

We have $\partial M^+ = \partial M^- = V_1^\perp$. Hence, we minimize the functional I restricted to M+M^+ and M−M^-.

Firstly, we consider the functionals I±=I|M±. Using Lemma 2.4, I± possesses the (PS)c conditions whenever $c < \min\{\int_\Omega T^+, \int_\Omega T^-\}$Therefore, we obtain that I± satisfies the (PS)c conditions with $c^\pm = \inf\{I^\pm(h) : h \in M^\pm\}$.

In this way, by using Lemma 2.2 for the functional I±, we obtain two critical points which we denote by h+0 and h−0 respectively. Thus, we have $c^+ = I^+(h_0^+) = \inf_{h \in M^+}\{I(h)\}$ and $c^- = I^-(h_0^-) = \inf_{h \in M^-}\{I(h)\}$.

Moreover, we affirm that h+0 and h−0 are nonzero critical points. To see this, from (E_4) and (E_6), we obtain that

$$I^{\pm}(h_0^{\pm}) \le I^{\pm}(t_1^{\pm}\Phi_1) = \int_\Omega F(x, t_1^{\pm}\Phi_1) < \min\left\{\int_\Omega T^+, \int_\Omega T^-\right\} \le 0,$$

and I restricted to $V_1^{\perp}$ is nonnegative. More specifically, given $\omega \in V_1^{\perp}$, using (L3-3) in Lemma 2.3, we have

$$I(\omega) \ge b_1 \ge 0. \tag{2.6}$$

Next, we prove that h+0 and h−0 are distinct. The proof of this affirmation is by contradiction. If $h_0^+ = h_0^-$, then $h_0^+ = h_0^- \in V_1^{\perp}$. Using (2.6), we obtain $I(h_0^+) < 0 \le I(h_0^+)$. Therefore, we have a contradiction. Consequently, we get $h_0^+ \ne h_0^-$. Thus problem (1.1) has at least two nontrivial solutions. Moreover, these solutions have negative energy.

3 PROOF OF MAIN THEOREMS

In this section, we prove Theorem 1.1, Theorem 1.2 and Theorem 1.3.

Proof of Theorem 1.1 From Lemma 2.4, the functional *I* satisfies the (PS)c conditions for some levels c∈R. Set E=V1⊕V⊥1, where V1=span{Φ1}. Using Lemma 2.3, we get that the functional *I* satisfies the saddle point geometry (see [12], Theorem 1.11). This implies that *I* has one critical point h1∈E$h_1 \in E$. Theorem 1.1 is proved. □

Proof of Theorem 1.2 From Lemma 2.5 and Lemma 2.6, we know that the functional *I* satisfies the geometric conditions of the mountain pass theorem. Moreover, the functional *I* satisfies the (PS)c conditions for all c≥0. Thus, we have a solution h2∈Egiven by the mountain pass theorem. Obviously, the solution h2h_2 satisfies I(h2)>0.

On the other hand, by Lemma 2.7, we get another solution h0h_0 and I(h0)<0. It follows that problem (1.1) has at least two nontrivial solutions. The proof is completed.

Proof of Theorem 1.3 Since the conditions (E_0), (E_3), (E_4) and (E_5) imply that Lemma 2.5 and Lemma 2.6 hold. Thus, we have one solution h2h_2 which satisfies I(h2)>0.

On the other hand, using Lemma 2.8, we obtain two distinct critical points h±0 such that I(h±0)<0. Therefore, we obtain that problem (1.1) has at least three nontrivial solutions. The proof is completed.

ACKNOWLEDGEMENTS

The authors express their sincere thanks to the referees for their valuable criticism of the manuscript and for helpful suggestions. This work was supported by the National Natural Science Foundation of China (11171220) and Shanghai Leading Academic Discipline Project (XTKX2012).

COMPETING INTERESTS

The authors declare that they have no competing interests.

AUTHORS' CONTRIBUTIONS

We declare that all authors collaborated and dedicated the same amount of time in order to perform this article.

REFERENCES

1. Garofalo N, Lanconelli E: Existence and nonexistence results for semilinear equations on the Heisenberg group. *Indiana Univ. Math. J.*1992, 41: 71–98.

2. Niu PC: Nonexistence for semilinear equations and systems in the Heisenberg group. *J. Math. Anal. Appl.* 1999, 240: 47–59.

3. *Silva, DA: Multiplicity of solutions for gradient systems under strong resonance at a the first eigenvalue. arXiv:1206.7097v1.* http://www.e-printatarXiv.org

4. Ahmad S, Lazer AC, Paul JL: Elementary critical point theory and perturbations of elliptic boundary value problems at resonance. *Indiana Univ. Math. J.* 1976, 25: 933–944.

5. Bartsch T, Li SJ: Critical point theory for asymptotically quadratic functionals and applications to problems with resonance. *Nonlinear Anal. TMA* 1997, 28: 419–441.

6. Furtado FE, De Paiva FO: Multiplicity of solutions for resonant elliptic systems. *J. Math. Anal. Appl.* 2006, 319: 435–449.

7. Landesman EM, Lazer AC: Nonlinear perturbations of linear elliptic boundary value problems at resonance. *J. Math. Mech.* 1969/1970, 19: 609–623.

8. Chang KC: Principal eigenvalue for weight in elliptic systems. *Nonlinear Anal.* 2001, 46: 419–433.

9. De Figueiredo DG: Positive solutions of semilinear elliptic problems. Lecture Note in Math. 957. In *Differential Equations.* Springer, Berlin; 1982.

10. Sara M: Infinitely many solutions of a semilinear problem for the Heisenberg Laplacian on the Hersenberg group. *Manuscr. Math.* 2005, 116: 357–384.

11. Willem M: *Minimax Theorems. Progress in Nonlinear Differential Equations and Their Applications.* Birkhäuser, Basel; 1996.

12. Silva EA: Linking theorems and applications to semilinear elliptic problems at resonance. *Nonlinear Anal.* 1991, 16: 455–477.

INDEX